Training Engineers for Innovation

Training Engineers for Innovation

Edited by

Denis Lemaître

iSTE

WILEY

First published 2018 in Great Britain and the United States by ISTE Ltd and John Wiley & Sons, Inc.

ISTE Ltd
27-37 St George's Road
London SW19 4EU
UK

www.iste.co.uk

John Wiley & Sons, Inc.
111 River Street
Hoboken, NJ 07030
USA

www.wiley.com

Library of Congress Control Number: 2018953437

British Library Cataloguing-in-Publication Data
A CIP record for this book is available from the British Library
ISBN 978-1-78630-356-1

MIX
Paper from responsible sources
FSC® C013604

Contents

Introduction

Why Train Engineers in Innovation?

The idea of training engineers in innovation seems obvious, just like training doctors in medicine or teachers in the art of teaching. Indeed, the word "engineer" is forged from the idea of novelty for producing something that does not exist in its natural state and that the human spirit invents thanks to its creative intelligence (*ingenium*). The engineer is the person who designs, and sometimes produces, innovative artifacts. This is what our engineers make today, whether it is an actual product, machines or abstract systems, models, tools, or even organizational methods, methods, processes, standards or representative forms of uses. The word innovation, as understood within our society, very nicely expresses this idea of producing something new and thus effectively articulates the core of the job of an engineer.

However, the link between engineering and innovation is no longer straightforward, as illustrated by the multiple injunctions addressed to schools to better train engineers in innovation, as well as the work undertaken by the same institutions to develop educational tools dedicated to this objective. There is therefore a gap between the new skills expected of the engineers in terms of innovation and the knowledge and know-how transmitted by the schools to their graduates, which up to now have allowed them to meet society's expectations. The question is where does this gap come from?

First, the new willingness to train engineers in innovation can appear to be a consequence of the recent evolution of the profession, under the effects

---

Introduction written by Denis LEMAÎTRE.

of the digital revolution and economic globalization. The period of the Taylorization of industrial activities and the rapid progress of technology have led, over the course of the 20th Century, to a standardization of tasks and to a specialization of the engineers in their areas of expertise. Today, since many of the technical engineering tasks (calculating, modeling, drawing, fabricating prototypes, testing, etc.) are largely automated, and since market logic leads to a focus on the needs of the client while integrating external constraints (social, legal, environmental, etc.), the work of engineers is considerably open and diversified. Engineers are thus expected to have the capacity to react rapidly to the needs of the market by proposing new artifacts (concrete or abstract), which are most often linked to new living and working conditions generated by these same technical advances, renamed with the more prestigious and modernist word "technology".

Second and more profoundly, the new injunction to innovation, in recent decades, seems to relegate to obscurity the traditional purposes of the figure of the engineer – linked since its appearance during the European Renaissance to the idea of progress (see Chapter 1) – which inspired the first training institutions. Progress, a word that symptomatically has fallen into disuse, covers purposes embracing society as a whole with respect to the liberation of the natural constraints weighing down on human life and the objective to create a better society on the foundations of reason. Compared to the idea of progress, the idea of innovation shortens both time and space, and tends to limit the activity of engineers to the immediate production of goods or services proposed on a given market, whether it be consumer goods, engineering services or even research products. It is no longer enough for young graduates to contribute via their work in a company, to strive to achieve better living conditions or technical advantages for communities, nations or humanity. They must also know how to transform technical advances into socioeconomic devices to introduce profitable technological products to the market via the creation of companies or new activities within diverse structures. This evolution deeply modifies the expectations that weigh on training systems and that are ever more restrained in their ability to adapt to these new conditions. The same market logic and injunction to innovation also applies to the higher education system.

What exactly is covered by the idea of innovation? It refers to an intellectual policy, a social position, a type of activity or a result of that activity; it resonates like a slogan and sometimes an exhortation, transporting ideological principles or different scientific paradigms; this notion is eminently polysemous. Just like a social and economic injunction, it has a positive connotation: in the common sense, innovation is always a good thing, which allows us to create employment, relaunch industry, meet new social needs, flourish, emancipate humanity, etc. Despite the astonishing semantic dispersion of the notion, it is this value that is highlighted by politicians, large institutions, speeches given by economic and social actors (including those of higher education and research), when they relay the injunction to innovation. How does this injunction resonate in the training systems for the engineers, who because of their job are on the first line to respond?

In order to provide an answer to this question, the research project *Innov'Ing 2020. Les ingénieurs et l'innovationmétiers, nouvelles formations* was undertaken and financed by the Agence nationale de la recherche (National Research Agency, ANR in French), a French public agency. Its objective was to study the contemporary changes in engineering training with regard to new expectations concerning innovation via an international consortium assembling researchers from different countries in Europe (France, Germany, Great Britain, Hungary) and countries referred to as emerging (Algeria, Morocco, Vietnam, Mongolia). Surveys were conducted in 2015 and 2016 on the training systems in the different countries, on the expectations expressed by the companies and on the pedagogical practices dedicated to innovation within the engineering training establishments (schools or universities). The current work assembles the contributions from researchers who have taken part in the surveys or in the final conference for the project and presents the results of this research, put into perspective and elaborated via theoretical reflections on engineering education. It aims to link the description of societal expectations and educational systems to concrete pedagogical solutions implemented, or made possible, within universities or engineering schools. The three-part organization is a product of this progression from the conceptions of innovation and the needs expressed, to the relations between the new skills, which are expected and the training systems, and the concrete pedagogical practices to respond them, either undertaken or projected.

The purpose of this work is not to offer a monolithic definition of innovation and an optimal pedagogical method to respond to it. It outlines the diversity of conceptions of innovation that are today in competition in the relationship between the technical and social dimensions of the engineering and training activities. The different chapters detail varied approaches, linked partly to the academic disciplines to which the authors belong (sociology, education sciences, management sciences, engineering sciences). Innovation is viewed turn by turn as a state of mind, a collection of processes, the results of a design activity and a purpose for action. Between the chapters some divergences appear in intellectual attitudes, for example taking the injunction to innovation for granted and looking for the means to prepare the students or deconstructing the injunction to understand what this idea conceals. There are also differences for the methods and the theoretical fields, between, for example, a management approach focusing on the conditions of entrepreneurship and a sociological approach focusing on the current innovation and social outcomes.

However, according to the results of the research carried out in the project, the different chapters converge to propose a rather broad conception of innovation, which translates into training objectives centered around decompartmentalization and the so-called complex approach, breaking with the disciplinary separation and the Cartesian approach to problem solving (i.e. breaking down a difficult question into many simple questions, solvable separately).

Indeed, the same intention runs through the different contributions to the work, which is refined toward innovation training that is inclusive or holistic, bringing together the technical, economic, social, environmental, ethical and legal dimensions of the problems. Some of the authors, teacher–researchers in human and social sciences within engineering schools, focus more specifically on showing the particular place that their disciplines must or can take, when linked with the technical disciplines, in the training of the engineers for innovation.

Chapter 1 reviews the transition from the idea of progress to that of innovation and considers its expression in the training of engineers. It allows us, thanks to historical and sociological analysis, to consider the question of innovation according to the situation and context, thus offering a conceptual framework for creating training practices.

Chapter 2 offers an original approach to thinking about the training of engineers in innovation around the perception of the event. This approach makes it possible to differently understand the need of mobilizing the idea of innovation to characterize the activities of the engineers, as well as the training requirements.

Chapter 3 offers an overview of innovation design from the viewpoint of economic activities and the company creation. It characterizes the expectations expressed by the corporate world for institutions that train engineers, thus presenting a framework in which to consider the pedagogical objectives.

Chapter 4 describes from a practical point of view the skills that the companies expect from engineers in terms of innovation. It presents, within a Hungarian context, an analysis of the forms of innovation that can be observed in the local companies and aligns them with the skills expected of graduate engineers.

Based on research concerning the relationship between innovation design and the educational practices in French schools, Chapter 5 highlights the contrasting approaches to innovator training. It connects the innovative practices to the construction of socioprofessionals identities characteristic of different types of innovators.

Chapter 6 presents a study conducted by the Agence française d'accréditation (French Accreditation Agency), Commission des titres d'ingénieur (Commission for Engineering Titles), on the theme of innovation and entrepreneurship. It outlines the innovation design defended by the CTI, before analyzing the way in which schools have assimilated the theme and applied the agencies' recommendations.

By questioning the relationship between the skills acquired in school and the skills acquired in the workplace in Hungary, Chapter 7 not only addresses the question of innovation, but also allows us to contextualize the question of innovation training in the relationship between training and expectations of the corporate world with regard to engineering.

Chapter 8 provides a practical example of innovation training within a German university. It puts forward, through investigations and an analysis of what currently exists, the way students can acquire the skills needed for innovation via dedicated devices.

Chapter 9 presents a device developed at a British university to train engineers in innovation. Based on a survey undertaken with the students, the training objectives given by the institution are confronted with the representations developed by the students.

Chapter 10 presents the results of a survey on innovation training methods in the French Engineering schools. It shows the complementarity of the different approaches observed between training on the development of technical applications and training on the analysis of societal needs.

Chapter 11 presents a policy of research action dedicated to innovation training within an engineering school in France. Based on two existing devices and using a skills matrix, it provides a framework for designing devices for innovation according to a holistic approach to problems.

The conclusion of this book is provided by André Grelon (EHESS). It allows us to put into perspective the question of innovation at the core of the training of engineers, with regard to the history and the potential developments to come.

PART 1

Innovation Design and Expectations toward Training

1

From Technological Innovation to "Situated" Innovation: Improving the Adaptation of Engineering Training to the Societal Challenges of the 21st Century

Since its appearance in Renaissance Europe as a social figure both recognized and promoted [LEM 03], the engineer has been seen as "someone who develops new techniques, signs of progress for society". At the time, the engineer was embodied by figures such as Leonard da Vinci, a figure situated in the imaginary as standing halfway between a creative genius and a sage. With the evolution of the profession, linked to advances in different technical domains and the implementation of standardized training, the image of the engineer has since evolved to embody a role, which is more technical and skilled than artistic. The engineer maintained a key social role throughout the modern period and up until the 19th Century: creating something new and thus producing progress. According to the dominant ideology of the time, the work of the engineer involves improving the living conditions of the population. Thus, there is the idea of a direct connection between technical innovations and social well-being, as well as a role, and even a moral position or duty of the engineer toward society, which is transmitted to them as part of their education.

Chapter written by Emmanuel CARDONA GIL, Linda GARDELLE and Brad TABAS.

From the second half of the 19th Century, the decline of the ideology of Progress has led to increased questioning of the existence of this link between innovation and social well-being, a link that was once thought to be both true and necessary. Gradually, the specific role taken by the engineer in contemporary societies has thus been challenged, and in its wake, the education of engineers has come into question. Our aim is therefore to consider what kind of innovators we need today, and how are these innovators are to be trained? The question discussed in this chapter is the role of the engineer in today's society, a role to be rethought in terms of the change in the historical–social paradigm that we have just described.

We observe that in the current situation, the effects of Taylorism and the market logic imposed by globalization sometimes confine the engineer to a role of technical expert. However, the new injunctions to innovation call into question the expected role of the engineer. This is the reason why we propose here to explore these expectations and the way in which training tries to answer them, the idea being to encompass the social roles of both the innovator and innovation. To do this, we start from the hypothesis that we are today emerging from a productivity paradigm and that we can from now on observe for ourselves the competing conceptions of innovation, corresponding to different purposes. We use as a basis for our investigation materials that have been gathered as part of the project Innov'ing 2020, including an overview of the training of engineers in Europe (France, Germany, United Kingdom, Hungary) as well as in some emerging Asian countries (Mongolia, Vietnam) and Maghreb (Algeria, Morocco). We have also considered speeches from different political and industrial actors with regard to their expectations in terms of innovation.

1.1. Progress and innovation

The definition of Progress as the idea that civilization has advanced in the past, is now advancing, and in all likelihood will continue to advance in the foreseeable future [NIS 1980], leads us to question the meaning of the word "advance". Since the beginning of the modern period, the notion of progress[1] has been closely linked to scientific and technical advances, which would have allowed us, according to a famous phrase by Descartes, to "rendre

1 Throughout the text, the word "Progress" written with a capital P designates this founding ideology of modernity.

comme maîtres et possesseurs de la Nature (make us masters and owners of Nature)" [DES 1637]. For others, progress implies not only the progressive liberation of humankind versus nature, but also the decrease in inequalities between human beings. For the most optimistic progressivists, such as Condorcet [CON 89], the advance of civilization implies the optimization of all aspects of human life: more power over nature, freedom, equality, reason, peace, health, etc. In this context, Progress is not simply a description of the accumulation of knowledge or wealth over time, but a veritable teleological law of nature and history. As in the liberal theory of the invisible hand (which was key for the development of the ideology of Progress), Progress produces a generalized well-being, according to which the efforts of isolated individuals transform the entire world for the better. Indeed, the dominant idea of technical progress has long been a liberal conception that assumes that all technical progress contributes directly to the progress of society by informing the individual pursuit of individual economic interests and hence the production of social well-being.

In this perspective, working to produce Progress can be seen as a moral duty, an obligation for the individual toward society or the human species as a whole.

1.1.1. *Progress and engineer training*

The idea of Progress and the social role that it confers to the engineer have played a paramount role in the representation of the engineer in the collective imagination of modern times, as well as in the thoughts linked to engineering education. William Barton Rogers, the founder of MIT, sought help for the founding of this school among "genuine friends of progress everywhere". He insisted on the fact that the institute, thanks to its progressive pedagogy, would provide "great public benefits" and demonstrate "the happy influence of scientific culture on the industry and civilization of nations". He based his remarks on the idea, very characteristically progressive, that "material prosperity and intellectual advancement are inseparably associated" [ROG 60]. We can see here the effects of the dimension of social engagement, inherent in the ideology of Progress, on the training of the engineer. From a pedagogical point of view, the implementation of this ideology translates into a turn toward modernity. The study of natural sciences, applied arts, modern languages, economic policy and constitutional[2] law is thus substituted for the more classical curriculum (ancient languages, history, literature, ethical and theoretical philosophy). Although generating the development of much scientific–technical knowledge of indisputable impact on society, this teaching model implemented by Rogers and his contemporaries is nonetheless today sharply criticized for its tendency to stifle the creativity of students, and thus to slow innovation [KIM 11]. In the end, the pedagogy of the 19th Century is a product of the idea of progress, and in this context, it is as much the scientific and technical discoveries as the "social" and discursive materials that fueled progress.

1.1.2. *Progress in crisis*

The idea of Progress has promised a great deal and has aroused a great deal of passion. It is undeniable that the last centuries have been synonymous with technicosocial advances that are generally beneficial in numerous domains – health, hygiene, education, material comfort – which continue to inspire some people to push still further the limits of what some have appointed "the infinite frontier" [BUS 45]. But the meanings and values attributed to these technical and social transformations underwent a clear turnaround in the middle of the 20th Century. Although there has always been criticism of progress – we can quote Rousseau, Thoreau, Neal Ludd or Burke – the idea that technical and economic progress does not always keep its promises has become more widespread. For example, the pursuit of "private benefits", although considered as a motor for economic growth, has been observed to lead to a disintegration of the social order, even toward the alienation of individuals and the dissolution of a sense of duty toward society. Thinkers such as Marx, and more recently Piketty in 2013, have demonstrated that the inegalitarian distribution of access to technological innovations and the financial means of liberalism were the reason for growth in inequalities, and therefore were the cause of a stagnation, even a regression of social progress. Being himself a progressive, Marx proposed political solutions to the problems of progress, which in turn created resistance and experienced failures that in themselves can be seen as arguments against progress. These failures have strongly marked the understanding of progress and have contributed to creating the so-called

2 This list provides approximate information on the subjects taught at MIT at the beginning of its existence.

postmodern vision of the world according to which we suffer from a "loss of historicity", according to which innovations are considered as not providing real social improvements, but as being mere market commodities that draw us even more profoundly into the historical paradigm of a terminal and hence non-progressive state of liberal capitalism [JAM 97].

But this postmodern point of view, as questionable as it is, remains one of the axes of the criticism of progress. Technical progress has also deceived. Some, such as Adorno and Horkheimer [ADO 17], saw in the progressive control of nature the progressive subjugation of men by technical means. In the same way, the horrors of modern war resulting from technical advances in domains such as chemistry or particle physics have revealed the dark side of technical progress, while at the same time killing the idea that modern humankind would somehow progress to a state of moral superiority relative to the human beings of previous centuries. Furthermore, there has been severe criticism of the discourses regarding progress by postcolonial thinkers like Mehta [MEH 99], who has shown that progress has very often been deployed as justification for imperialist and colonial projects, and that it has been inseparable from a lack of respect toward local ways of thinking and doing. Even more recently, Brynjolfsson and McAfee [BRY 15] have warned about the fact that innovations in automation within factories, made possible by advancements in robotics, risk producing mass unemployment and an increase in other forms of social harm. From an ecological point of view, technical progress is also ambiguous. First, the ecological perspective demonstrates the impossibility of endless growth (which was at the core of the progressive vision), because natural resources are not inexhaustible and the planet has determinable physical limits [MED 04]. Likewise, some technical advances such as the discovery of the steam engine, which had been the pride of the progressists, have revealed themselves to be the cause of catastrophes or threats such as climate change. In this respect, society in general has experienced what the philosopher Hans Blumenberg called an "Unbehagen am Fortschritt" (discomfort with progress) [BLU 15], meaning that changes due to technical progress have been understood to be not always wholly desirable, even if at this point that have come to seem almost inevitable. In sum, over the course of the 20th Century, the very idea of progress as a universal project involving all of humanity has been widely called into question.

1.2. Rethinking progress

Technical and economic progress continues to boom to such an extent that some people worry regarding the acceleration of the changes that it generates [ROS 05]. Very few people today believe that Progress has fulfilled its initial social promises. The reasons for the failure of this ideology of Progress in its classic version, regarding the improvement of the living conditions for populations, is mainly due to the absence of theorization about the practical means for the improvement of society. This is largely a question regarding the means by which this was supposed to happen. If we want to generate social progress starting from economic and technical progress, should it not be considered that social, technical and economic dimensions must be linked? Relying on sociotechnical innovations, in a more encompassing project for society, should make it possible to overcome the limits that we see today.

1.2.1. *The engineer, innovation and crisis in Progress*

Confronted with a new necessity, the engineer, unlike individuals in general, now has difficulties when thinking about the purposes of the work. As individuals, engineers are confronted with ethical dilemmas regarding their activities that sometimes overwhelm them.

The recent increase in interest given to the notion of innovation is an indicator of the disappearance of the sociotechnical ambition for progress. This is shown by research that we have performed on Google NGram[3] via an analysis that takes into account the frequency with which words are used in almost all of the texts which have been digitized, the English terms progress and innovation were used widely in the 15th Century, but the former quickly became the most frequently used, a trend which lasted up to the crisis of the ideology of Progress, before reversing during the 20th Century. Generally, it can be said that for a creation to claim progress, it is essential that it should be directed toward a purpose that must be linked to the improvement of human society. With respect to innovation, the ambition is lower: for a creation to be considered an innovation, it must be new, but this novelty is not necessarily linked to a larger teleological and social vision. If innovations that meet social needs do exist, an innovation can be considered

3 This research was carried out in October 2017 on Google Ngram, with numerous entries of the terms "progress" and "innovation". There were between 1,500 and 2,000 entries.

as one even if it meets no existing need, and we all frequently find celebrated innovations that create new needs. In this context, the increasing interest in innovation is explained partly by the expansion of a market and individualistic vision of the world. Thus, innovations are often the origin of market growth built upon a short-term pursuit of profit and at the detriment of long-term goals such as sustainability.

Many schools and governments are today adopting positions halfway between progressive and postprogressive visions, pursuing policies of technical and economic innovation, but without any real effort to improve their profitability from a social point of view. Our investigations have shown that from the western tip of the Eurasian continent up to its extreme East, speeches which embody this ideology are given during major meetings about engineering training, and also in the highest levels of discussion about engineering training in the field, in schools, with students. Quoting, for example, the important meetings represented by the Global Engineering Education Leader Conference, which brings together each year the directors of the largest establishments and agencies' concerned by engineering training in the world (representatives of American and Asian universities, the International Federation of Engineering Education Society's, the Global Engineering Dean's Council, of the UNESCO International Centre for Higher Education Innovation, the Commission des titres d'ingénieur (Commission for engineering titles, etc.). During this conference, which took place in Shenzhen in China on December 5–7, 2016, the remarks of certain business representatives were very clear: Innovation equates to competitiveness with the ultimate aim of profit. The opening speech of the conference, made by the executive president of a major Chinese company, could be summarized as marketing praise for "to appropriate technology to market", "innovation has to meet market needs" and "you should do everything in a business logic"[4]. Such conceptions also exist among leaders of institutions engaged in the training of engineers. For example, the president of SUSTech (Southern University of Science and Technology in China) gave a detailed picture of what innovation represents for his institution: a collection of precise and ambitious figures organized in a table (inventory and objectives in terms of number of patents, number of theses, profitability in dollars, etc.). Such speeches can also be found in engineering schools and universities in Europe. They perfectly reflect that the dominant ideology of innovation is orientated toward technologism and business

4 These remarks, like the previous, have been collected by the authors.

profits rather than toward meeting human needs and the development of human society. Other visions also exist regarding innovation and the purpose of engineering training. Ethics, social entrepreneurship, sustainable development and social responsibility are issues that are increasingly raised.

1.2.2. *The technical and economic paradigm and innovator training*

The design of progress reduced to technical and economic objectives has a strong impact on the training of engineers and innovators. In most countries, higher education has transformed since the 1990s, as a result of economic globalization and political will. Since the European Union was given, in 2000, the objective of becoming the most competitive knowledge-based economy in the world, with training and industrial dynamism as its main priorities. The EU aims to encourage innovation and scientific excellence. In order to harmonize higher education in the European area, profound upheavals have taken place in the higher education over the last two decades with the generalization of a division in three cycles: BA, MA, PhD (translated in French by licence, master, doctorate). Beyond the political will to develop these measures, it is the economic and financial globalization as well as the objective to develop the market, which has driven these changes in the world of higher education. The objective was to encourage the free movement of students, liberalization, the competitive diversification of the programs available and the move toward a "global harmonization of standards and academic titles". This was also supposed to stimulate the competition, which comes via the search for attractiveness, with students becoming clients [LAV 02, LEB 06, LEC 11]. This academic globalization has resulted in global distribution of pedagogical and academic models, thought patterns and new excellency standards [MAZ 09], often in line with the technical and economic logic of the liberal progressivism where the improvement of collective well-being comes second to individual self-interests.

With respect to engineering training, the current evolution of higher education and the consequences of internationalization have slightly erased the differences that existed between the different initial training models and thus moved toward a standardized model. Most of the representatives of higher education establishments met during our investigations did not have a clear way of presenting the means implemented to train engineers capable of innovating, but all said that they have thought about these questions and are trying to develop educational programs with this in mind. These operations are conducted at different levels of thinking and detachment: many establishments do not have the means or the objective to question everything; they sometimes remain more concerned with the very clear expectations of the companies or settle for pedagogical experiences resulting from trends. Others develop diverse and ambitious devices (refer to Chapter 6).

According to our investigations with prescribers of training within the engineering schools (pedagogical managers, training directors, etc.) regarding their representations of innovation, what seems to create a consensus, and this is corroborated by the specialist literature [ALT 00, BES 13, LIU 15], is the idea that to become an engineer capable of innovating, learning to be autonomous and proactive is essential. If the curricula are not the same, in the sense that the goals regarding the engineer that we wish to train and the organization for the different training differ, pedagogical practices tend to evolve, slowly and at different degrees, toward a common model: give more autonomy to students, use attractive digital pedagogies, open up internationally, etc. For this, as shown by our investigation, the work by individual and collective projects are put forward as exemplary, whether in French, Moroccan or Algerian engineering schools, Hungarian colleges or in some Vietnamese training where we introduced, for example, end of study projects. Active pedagogies and creating accountability for the students, like everything that will encourage leadership, are also sought in different training schemes by miscellaneous devices, notably in Great Britain. Digital pedagogies are also mentioned by our contacts (MOOC, digital platforms, etc.). The link with research is also frequently highlighted. Some countries are ahead on this aspect, for example in Germany where the links are very strong between training and research, and where the PhD is better valued than elsewhere. Entrepreneurship is also valued and put forward as a generator of innovation. Thus, incubators and spaces to support business creation are created in most countries, with diverse means. More broadly, the link created with companies during the training is deemed essential. Germany is a role model for this. Most of our contacts have also underlined the importance of non-technical disciplines in the curricula, even if the choice of the materials is often limited to management, in a "neopragmatic" way, which aims to equip the students with the tools to successfully integrate in the corporate world [LEM 14]. Also, transverse projects and interdisciplinarity have been evoked as a significant stimulus for our contacts. Mixing the training of engineers with those of designers makes them work with sociologists or artists, which will allow from this point of view to encourage openness, promoting reference frameworks, which can only be a vector for creativity. Finally, international mobility has been underlined by some of our contacts as a way for students to be more open and more agile in thinking, qualities considered as at the origin of the innovation process. Resetting one's bearings, leaving one's daily framework, seeing other ways of doing things, being confronted with difference, questioning oneself, feeding oneself with other ways to see and to think, taking a step back from its society. All these are means to grow and become more open minded.

1.3. Rethinking innovation and the innovator

Are these attempts to train innovators beyond the scope imposed by the ideology of innovation and the technical and economic paradigm? Can we encompass the practices that align more with a vision about sociotechnical innovation? How to establish the difference between independence of thought and independence to act that allows innovators to provide sociotechnical solutions to their companies, and another, more orientated toward the person, implying an acceleration of the neoliberal logic? An attempt to respond involves suggesting to students the importance of thought and the problems with regard to innovation, made possible notably with regard to the lessons in human and social sciences.

The devices, in Europe or elsewhere, are presented as greatly participating in the demand for the new skills required for project management, the adaptation to economic and financial imperatives, and/or sometimes to work in multicultural companies open to the world. The mobilization of human and social sciences in the curriculum, often reduced to their purely instrumental aspect (accounting, marketing, business law, etc.), shows the logic that they must serve: providing young professionals that are able to integrate rapidly and easily into the corporate world. According to this vision, the training of engineers as innovators must meet the conditions of opening of the market and the liberalization of the economy, serving first of all the companies and meeting the standards of economic globalization based on a cost/benefit calculation in the short term via the creation of new needs. The vision of the modern engineer is to achieve technical progress in society to improve the living conditions of the population; in this context, a long-term vision that seeks to respond to the needs of all is sometimes neglected. We can also overlook, or at least underexploit, the even older vision of the engineer as a creative artist who is able to apply technical know-how as well as aesthetic judgment. Sometimes the much larger reach that the human and social sciences can have in their reflexive and cultural dimension can be neglected, given that these disciplines allow young engineers to work in companies by understanding the practical realities, but also concerned with the multiple dimensions of a project, have a sociotechnical vision for the fruits of their work and have a much more inclusive approach to the world that surrounds them.

The point is not to go back in time and remake progress with all its contradictions, but to promote a way of thinking about innovation, and thereafter the training of innovators, more closely linked to social values. Rather than seeing "innovations" as disconnected from all social value, it is thus important that the engineers develop an awareness of their social responsibility, such that they do not merely develop innovations because they are new, or even a source of wealth, but because they bring about positive changes both from a technical and social viewpoint. The complexity of modern societies requires not only specializations in advanced technical and scientific domains, but also a better understanding of the world and the forthcoming challenges [AKA 03, JAM 11] in order to provide innovations with a positive social impact.

It is important to add that there is a willingness to train such engineers. Many leaders of training agencies are concerned about enlarging the vision of innovation within engineering training to underline the necessity for these future professionals to be aware of major societal and environmental issues. For example, Seeram Ramakrishna, founder of the Global Engineering Dean's Council and director of the Faculty of Engineering at the University of Singapore, stated that "the 21st Century requires training engineers capable of meeting modern day challenges in terms of socioeconomic development, locally, regionally and globally. Creativity and innovation must serve this objective". Thus, different conceptions oppose each other. In France, work undertaken by researchers on human and social training of engineers are a notable influence. For several years, CTI management has expressed the importance for future engineers to generate a reflexive capacity and emotional distance, which will allow engineers to be more farsighted on the different aspects of technology in society and on the evolution of society itself [GUI 06, LEM 03]. Business leaders also want the curricula to evolve toward putting in place a new learning device encouraging emotional distance on problems. Some are aware that market globalization and challenges raised by industrialization require engineers open to social and economic dimensions of the problems and adaptable to different professional cultures. Within training agencies, it may not be a majority, but some Pedagogical managers are aware of this requirement for openness.

1.4. Training for a "situated" innovation

One of the main criticisms aimed at the ideology of progress lies in its abstraction. In fact, it was thought for a long time that there were necessary and abstract links between technical progress and improvements in living conditions, so no one truly worried about the creation of these links. To link technical progress with society, attention must be paid to details and specificities. We must ask ourselves what type of innovators are needed and if the programs available are able to create such innovators. In this context, we must question whether training carried out in compliance with the standards imposed by the market might be a risk of creating innovators that are blind to societal needs?

The empirical studies that we have conducted in several European and emerging countries show that injunctions to innovation are widespread at an international level. Everywhere, we have heard that innovation is an objective for the government in power, even if the measures that are instituted are not the same. If we take, for example, two countries in Maghreb, Morocco and Algeria, the differences are considerable. For one, significant resources have been deployed for several years and have been successful; for the other one, the speeches have not been accompanied by solid and coherent initiatives. Ultra-modern techno poles are dynamic in their relations with training agencies and are completely internationalized in Morocco, in sharp contrast with the Techno park of Sidi Abdallah in Algeria, far from Algiers, training establishments and economic life [GAR 15]. The meaning given to innovation differs quite clearly from one country to another. Within the same country, those involved interpret differently the importance to innovate. In short, the rhetoric about innovation collected by the different people and institutions met is rich and plural. Different points of view oppose and compete. Empirical studies carried out in engineering training establishments as well as among captains of industry and other leaders show divergent views on the conceptions of what an innovative engineer should be.

Some people highlight technological innovation, others innovation in uses, others what is linked to uses, innovation in the interpretation and the adaptation of standards, objects and practices to new contexts. For example, we can mention what we have observed for some actors in Morocco, who intended to adapt everything coming from elsewhere to the context and uses of Africa; or what politicians in Mongolia intend to achieve by "mongolizing" techniques to adapt them to their specific environmental and cultural environment.

If we take the example of Morocco, which significantly highlights the knowledge-based economy (putting in place numerous large-scale plans or projects, recent creation of dynamic techno poles, etc.), innovation is defined as one of the main priorities and the first source of competitiveness by the Moroccan government [DJE 12]. In this context, a consensus seems to emerge from speeches of the various actors interviewed, presenting a large part of the innovation system as being linked to the way in which Morocco can take advantage of the global knowledge storage, to appropriate it and distribute it by adapting it. It is not necessary to reinvent everything, but to adapt the services, products or the methods coming from elsewhere to the conditions of Morocco, even Africa. "Morocco should innovate as a gateway toward Francophone Africa. To innovate is to adapt things, a hub toward all of Francophone Africa. This would be a great challenge for Morocco", declared an engineer responsible for collaborative solutions at Cisco in Casablanca. This strategic positioning, vision in the direction of Africa, is also highlighted by Moroccan decision makers at all levels. This message is also highlighted by the techno poles to attract foreign companies.

This idea of innovation by adaptation can also be found in Mongolia, a country with a long tradition of nomadic pastoralism, from which the government intends today to create an environment encouraging knowledge-based economy, the development of innovation in society, economics, politics, science and technology. In this context, i.e. confronted with the challenges of changes in the way of life or even population growth, the leaders must be careful to take into account the experiences of other countries to innovate and experiment [GAR 16]. Also, governors, as well as the ordinary Mongolian citizens at their level, do not fail to observe what is developing in the other countries of the world and to take it as an example in order to adapt existing technics according to their specific environment (extreme climatic conditions, country with one of the lowest population density in the world, etc.). To do this, the government relies heavily on its student diaspora and therefore encourages student mobility in foreign universities, such as the development of international partnerships and the installation of foreign training establishments in Mongolia.

This innovation design in the interpretation and the adaptation of standards, objects and applying practices to new contexts is in line with the definition of the innovator proposed by Liu [LIU 15]: the innovator is somebody who updates tradition, somebody who is capable of making tradition fit for purpose, to make sense of the past in the context of the future. And for this, the innovator must have a global vision of the past, present and future of society, but also an understanding of the different operating levels of the groups and organizations: links between the micro (group level), meso (organization) and macro (society). In this context, some highlight the importance of training about transversality and the continued link, which encourages intuition as opposed to the segmentation of various knowledge. Although renewed, we can find in this perspective the spirit of progress.

From this perspective, a certain approach to innovation can be characterized as "situated", in the sense where it relates to a particular national context and its specificity [CAR 17]. Taking the example of Vietnam, which has seen the issue of training become central since the liberalization of the economy in 1986, and where the government today calls for reform of higher education to put it expressly at the service of industrial and economic development, development can for some people be a real philosophy of life. During an interview, a company director conceptualized this approach using the title "Vietnam Dreams" in the context of economic development directly linked with tradition. In this perspective, it is above all not forgetting the past to build the future, which the person interviewed denounces as the dominant position today: "we strive to forget the past to create the future", but on the contrary, we rely on our cultural heritage as a source of development that makes sense. Hence, harsh criticism about the way in which change can be driven can have the following result: "people are mistaken about development, they don't have common sense (…) it is often seen just as a cost/benefit calculation, whilst it is essential to understand things in their entirety". The use of these techniques must therefore, in this perspective, continue with clear targets, notably vis-a-vis society. It must also be perceived as a means rather than as an end: "technology makes no sense in itself [...] technology is not a finality, it is its use which must make sense". Also, people close to this logic argue in favor of an education that is not dehumanized, but which on the contrary must make engineers aware of the global issues they could address, accomplishing aims that we could describe as humanistic. Taking all of this into account, and recognizing that the task is made more and more difficult by the growing corporatization of the university, the role of the educators and educational institutions is to create for their students the conditions for thought.

1.5. Conclusion

The results of the investigation presented in this chapter show that engineering training must be in line with the needs of the population and the significant challenges of the present. For engineers, a major objective is to become aware that via their activity, they serve a greater cause that goes far beyond the individual or the market. In this context, the idea of an all encompassing sociotechnical approach, based on the needs of the population and not exclusively on high-tech advances, can provide an interesting base for rethinking engineering training about innovation. It is not a total reformulation of the knowledge already possessed, but a reorientation of the teaching practices and a revision of educational expectations, creating an awakening of the social conscience in the engineers coupled with a broader understanding of the way in which social issues and technical objects interlock. Engineering training today must aim to provide a much more precise understanding of the sociotechnical means necessary to perform this role, but it must also allow a new vision of the engineer to emerge. The Latin term genius, etymologically at the source of the words "engineer" and "genius", refers to the creative power shared by artists and inventors. If in theories of Progress, it is very much this genius of human spirit, which is considered as the motor for progress; for several decades, engineering training has sometimes devalued this creativity, so appreciated in the work of the first engineers of the modern era. However, given the current injunction to innovate, one wonders if this creative capacity is not too often ignored in favor of the cost/benefit analysis and a focus on advanced technologies.

Our work aims to participate in thinking about design engineering that also includes its social mission. We also defend in this chapter the theory of the need for situated innovation that, without being substituted for technological innovation, makes it possible to train engineers to have a more reflexive position, even to understand the issues specific to their society and therefore to respond to them.

From a more general point of view, this paradigm shift, which we have identified in our research, by which the activities of an engineer are performed for the benefit of society, deserves to be analyzed in terms of the tensions created with the techno-economic model. It remains to be seen if we will be able complete this shift, and whether the sociotechnical model will supplant the technoeconomic model in the years to come. Perhaps we will rediscover the interest in what was meant by "engineer" in the past (the engineer as a provider of progress and innovation) as technological innovations emerge, which are better able to respond to current societal challenges.

1.6. References

[ADO 17] ADORNO T., HORKHEIMER M., *Dialektik der Aufklarung*, De Gruyter, Hamburg, 2017.

[ALT 00] ALTER N., *L'innovation ordinaire*, PUF, Paris, 2000.

[BES 13] BESANÇON M., LUBART T., BARBOT B., "Creative giftedness and educational opportunities", *Educational and Child Psychology*, vol. 30, no. 2, pp. 79–88, 2013.

[BLU 15] BLUMENBERGH H., *Schriften zur Technik*, Suhrkamp, Frankfurt, 2015.

[BRY 15] BRYNJOLFSSON E., MCAFEE A., *The Second Machine Age: Work, Progress, and Prosperity in a Time of Brilliant Technologies*, Norton, New York, 2015.

[BUS 45] BUSH V., Science the endless frontier: a report to the president, United States Government Printing Office, Washington, 1945.

[CAR 17] CARDONA GIL E., LEMAÎTRE D., "Entre mondialisation et développement local, quels modèles pour la formation des ingénieurs au Vietnam ?", *Cahiers de la recherche sur l'éducation et les savoirs*, no. 16, 2017.

[CON 89] CONDORCET M., *Esquisse d'un tableau historique des progrès de l'esprit humain, suivi de "Fragment sur l'Atlantide"*, Flammarion, Paris, 1989.

[DES 70] DESCARTES R., *Discours de la méthode*, Le Livre de Poche, Paris, 1970.

[DID 99] DIDIER C., *Pour un questionnement éthique des choix techniques. Une ouverture dans la formation des ingénieurs*, Éditions Charles Léopold Mayer, Paris, 1999.

[DJE 12] DJEFLAT A., *Les efforts du Maroc dans l'économie fondée sur la Connaissance*, CMI, Marseille, 2012.

[GAR 15] GARDELLE L., DROFF J., NAFA A., "L'expérience technopolitaine en Algérie: enjeux et perspectives de la diaspora pour l'attractivité territoriale", *Méditerranée: revue géographique des pays méditerranéens*, vol. 124, pp. 65–73, 2015.

[GAR 17] GARDELLE L., CARDONA GIL E., BENGUERNA M. *et al.*, "The spread of European Models of Engineering Education: the challenges faced in emerging countries", *European Journal of Engineering Education*, vol. 42, no. 2, pp. 203–218, 2017.

[GUI 06] GUICHARD R., SERVEL L., "Qui sont les innovateurs ?", *Sociétal*, no. 3, pp. 26–31, 2006.

[JAM 87] JAMESON F., *Postmodernism or the Logic of Late Capitalism*, Duke University Press, Durham, 1987.

[KIM 11] KIM K. H., "The creativity crisis: the decrease in creative thinking scores on the Torrance tests of creative thinking", *Creativity Research Journal*, vol. 23, pp. 285–295, 2011.

[LAV 02] LAVAL C., WEBER L. (eds), *Le nouvel ordre éducatif mondial*, Nouveaux regards/Syllepse, Paris, 2002.

[LEB 06] LEBEAU Y., "Pressions sur l'enseignement supérieur au Nord et au Sud", *Cahiers de la recherche sur l'éducation et les savoirs*, vol. 5, pp. 7–14, 2006.

[LEC 11] LECLERC-OLIVE M., SCARFO-GHELLAB G., WAGNER A.-C., *Les mondes universitaires face au marché*, Karthala, Paris, 2011.

[LEM 03] LEMAÎTRE D., *La formation humaine des ingénieurs*, PUF, Paris, 2003.

[LEM 14] LEMAÎTRE D., "Quelles finalités pour les sciences humaines et sociales dans les formations d'ingénieurs ?", *Kalim*, vol. 2, pp. 27–40, 2014.

[LIU 15] LIU T., "Quelles pédagogies pour former des innovateurs ?", *Actes du VIIIe colloque Questions de pédagogie dans l'enseignement supérieur*, Brest, France, June 2015.

[MAZ 09] MAZELLA S., *La mondialisation étudiante*, IRMC-Karthala, Paris, 2009.

[MEA 04] MEADOWS D., RANDER J., MEADOWS D., *Limits to Growth. The 30-Year Update*, Chelsea Green Publishing, New York, 2004.

[MEH 99] MEHTA U., *Liberalism and Empire*, University of Chicago Press, Chicago, 1999.

[NIS 80] NISBIT R., *History of the Idea of Progress*, Transaction, New York, 1980.

[PIC 13] PICKETTY T., *Le Capital au XXIe siècle*, Le Seuil, Paris, 2013.

[ROG 60] ROGERS W., *Object and Plan of an Institute of Technology*, John Wilson, Boston, 1860.

[ROS 05] ROSA H., *Beschleunigung. Die Veränderung der Zeitstrukturen in der Moderne*, Suhrkamp, Frankfurt, 2005.

2

Responding to an Event: Innovation of the Contemporary Engineer?

Contemporary capitalism is characterized by the emergence or accentuation of new sources and forms of innovation. We consider innovation, in a very general and preliminary way, as a dynamic significantly transforming the value created via the use of a technical device, whether this device targets end users ("product" innovations) or concerns the organization of labor and production practices (organizational and managerial innovations), or even institutions [LE 06].

However, in contemporary capitalism, innovation seems to have been initiated and developed starting from new "sources". Many works have tried to shed light on the question of innovation sources to further enlighten the innovative process, which is central to the competitiveness of contemporary organizations [CHE 03, CHR 97, DRU 85]. Innovation may arise from localized initiatives of actors and may emerge through diffuse and random events, which may be a result of incentives or, more fundamentally, from unexpected events, which may be individual or collective. The situations or contexts conducive to the emergence of innovations are extremely heterogeneous. In this chapter, we will attempt to reformulate the process of innovation based on the dynamics of contemporary capitalism, involving uncertainty, instability and unpredictability. This link between uncertainty and innovation has already been problematized [ALT 93]; it shows that innovation today has a more radically unpredictable character, no longer emerging or starting from controlled initiatives, but occurring in response to

---

Chapter written by Frédéric HUET, Hugues CHOPLIN, Isabelle CAILLEAU and Pierre STEINER.

the manifestations of instability and transformation that we will refer to as "events". This uncertainty involves the engineer in a new relationship with innovation, which may prove to be an uncomfortable one when the position is no longer of involved with being a "pilot" or "manager" of innovation, but rather that of an actor capable of transforming these disruptions into potential innovations.

These new forms of innovation invite a conceptual renewal that we perform by insisting on the notion of an event. We believe this leads in turn to a reformulation of the characteristics of contemporary innovation, to redefining the position and the "action" possible by the engineer and to considering events from a positive and constructive stance: a trigger for new sources of value creation, beyond their destabilizing even destructive character. We will attempt to discuss this hypothesis about innovation as a response to an event. We will present the use of a study of three situations involving engineers facing an event. These situations, observed using an investigative pedagogical approach with engineers, illustrate this "event base" of engineering activity and offer empirical material for processing the type of response given by engineers.

2.1. From contemporary capitalism to innovation situations

2.1.1. *The dynamics of contemporary capitalism...*

While industrial capitalism was founded on strong regulation and strong control of productive activity, including the production of innovation[1], contemporary capitalism is characterized by great instability (market fluidity, recomposition of value chains, the emergence of disruptive technologies, etc.) and the associated uncertainty[2]. Several trends in contemporary capitalism combine and lead to the instability of markets and organizations. We will mention three of them in this chapter: the production

1 For engineering training, the lessons for innovation management and innovative project management are all tangible manifestations of this demand for innovative processes that can be manipulated or controlled.

2 A hypothesis regarding this contemporary instability could be attributed to the contemporary transition period between two stages of capitalism: from industrial capitalism having developed until the *Trente Glorieuses* (1945–1975 in France) its modes of regulation to cognitive capitalism or an economy based on knowledge, again largely unachieved and having not found its modes of regulation [HUE 17].

of relational value, the financialization of organizations and reticular production.

Industrial logic was based on the production of value anchored in the process of material transformation. In other words, the added value produced was very largely crystallized in the material objects produced and exchanged. Contemporary capitalism displaces the sources of added value, now largely created outside strict production activities (for the transformation of material). Production now integrates a service requirement that results in a requirement not to offer material goods to a global and anonymous demand, but contextualized solutions responding to unique problems[3] jointly constructed between producers and users [HUE 12, MOA 09]. This servicization involves the mobilization of intangible and relational assets (reputation, confidence, creativity, reactivity, etc.) where industrial requirements favor material assets and their accumulation. Yet, these intangible assets are more volatile than material assets, subject to obsolescence or anticipatable depreciation. It is thus in the capacity of organizations to produce and maintain relations with a multitude of stakeholders that new sources of competitiveness can be found.

The second trend is the financialization of activity and new relationships formed between the financial and productive spheres. Previously, shareholders (institutional investors among others) were largely subordinate to productive technostructure, but now take a much more active role in the governance of a company and import their conventions and performance criteria to direct productive activity, at the same time redefining the relations of those in power, and the way in which this power is exercised within the organization [SEG 12, WEI 10]. Conflict can emerge from the different temporalities induced by the cohabitation of financial and non-financial, strategic and productive requirements[4]. Financial performance and financial market dynamics are based on a logic of opinion and attention that results in an evolution of the activity and shows results at an accelerated pace, which can interfere with the longer temporality of productive activity.

3 The situation of consultant engineer (situation 3) that will be described further can be viewed as a direct implication of this renewal service of the activity.

4 The situation relative to the unexpected closure of a production site (situation 2, see Figure 2.1) can be read as resulting from a distance created, by these new forms of governance, between the "places" of decision and the places of production.

Finally, the third trend that we will explain is that of reticular production. Diverse evolutions (technological, financial or commercial) have lead to a strategic repositioning of productive organizations toward specialization in their core competencies. The consequence of these dominant strategies is the progressive deconstruction of the Chandlerian corporation, the organization mainstay for industrial capitalism. The shift is from an internal organization offering a very high degree of integration, and total control of the productive skills required for a productive organization, to a productive organization that is part of a network, within which skills are distributed between a multitude of interdependent stakeholders[5]. Production and skills are no longer concentrated in the company, forcing the latter to enter into a productive ecosystem, to develop, maintain and update interorganizational links (beyond "meager" market transactions) to enhance its contribution to the productive process [MOU 08].

To recap, contemporary capitalism is characterized by new interdependencies that form the basis of its complexity and instability. Although industrial capitalism is characterized by strong "acceptance" and control within a large company, contemporary capitalism is much more open, and is no longer "controlled" from the inside of the organization but guided by interorganizational reconfigurations with the instability that this implies.

2.1.2. *... to new situations of innovation*

Several manifestations of this contemporary capitalism dynamic can be observed in the company, starting from problematic situations [DEW 93] in which engineers, notably, must act. We can associate or regroup[6] these problematic situations according to three main categories of contemporary

5 The collaboration described in situation 1, between two teams, one French and the other Indian, is also considered as an illustration of the passage to cooperative production, in the network.

6 The three large categories of problematic situations present in companies are here proposed starting from the accumulation of information on the activities of French apprentice engineers at the université de technologie de Compiègne, to which it has been requested to identify (with regard their university training) the problematic situations in which they could be engaged with. This categorization has no pretense to a statistic representativeness or analytical robustness, it allows rather to seize and to illustrate the manifestation of the current organizational instability.

(inter)organizational phenomena (or metasituations), which are more or less the direct implications of the trends that we have described above.

First, the management of the boundaries of the activity and the organization complicate the engineer's activities[7]. Previously confined to a local and internal zone, activity is today deployed outside of the boundaries of the organization (if we can trace its boundaries), which implies management of distant relations – physical and cognitive distance – with different partners, whether they are clients, suppliers or competitors... Management becomes the integration of the specifics of these new collaborative activities within production and innovation networks: reconciling divergent interests, heterogeneous work practices, communication constraints, etc.

Situation 1: The development of a CRM tool with an Indian partner[8]

The case of the development of a Consumer Relationship Management (CRM) tool with regard to a Franco-Indian partnership allows us to highlight this first problematic metasituation linked to the boundaries of activity. The company SI (a multinational specialized in offering energy management solutions) wished to unify the CRM tool used by all of its subsidiaries throughout the world. In order to decrease its development costs, SI management decided to subcontract development/coding tasks to an Indian partner, SF, which provides its own technical solutions. It was therefore up to an engineer within SI to ensure the transmission of the different elements necessary for the development of the site (specifications, functional specifications, etc.) and to copilot the project with an Indian counterpart. Despite some coordination problems at project launch, it took place without any real difficulty. But after the first managerial meeting between

7 Even if these metasituations are not exclusively from one of the three mutations presented previously, we see that this porosity of organizational borders is the translation of the passage to the logic of production and innovation distributed in the network. Likewise, the difficulties linked to a decisional process can be linked to the financialization of productive activity. Furthermore, the requirement for improvization is partly nourished from this tendency for servicization.

8 The information on these situations has been collected via interviews with practicing engineers. These interviews were explicitly oriented toward the "problematic situations", which the engineers could be confronted with during their career. They are not representative or exhaustive regarding the activity of an engineer; they are rather a focus on the situations, which have marked those interviewed and on their response. The limit of this collection device means that a longitudinal follow-up has not been performed and that the consequences in the term longer have not been questioned, which is limiting to rigorously prepare a discussion on innovation, inseparable from a dynamic of diffusion and socialization.

the French and Indian teams, the situation degraded significantly. Disagreements were frequent, and the Indian team abandoned any collaborative attitude. The cause of this degradation was identified by the French Engineer/project manager as intercultural: the Indian coordinator stopped all communication with his French counterpart when he saw that it was a woman.

This event badly impacted the progress of the project and required a response to go beyond this dispute and allow the creation of the CRM tool as initially envisaged. This response, which we detail after this, produced only a very limited result, because although the project has continued, the solution developed remains unsuitable and resulted in a project relaunch to completely respond to the client requirements. Neither a success nor a failure, the project seems to have stalled, causing delay and underperformance with regard to the initial expectations.

Box 2.1. *Development of a CRM tool with an Indian partner*

Following on from that, governance and the decision-making processes within a contemporary organization often result in contradictions or constraints to be integrated in the activity. Engineers, as intermediaries, often find themselves in a position that requires them to transform strategic objectives into operational ones and to find a compromise between these strategic requirements and the constraints and practices on the ground. Also, reconciling the short-term requirements without forgetting the longer term issues and allowing the contributions to coexist in a project mode without overcoming the imperatives of the organizational structure are all expectations and requirements that make carrying out the activity much more problematic.

Situation 2: The closure of a production site

As a method and process engineer within the FA group (automobile parts maker), S. was promoted to the role of quality manager at a Brazilian site of the group. To develop its activity in Brazil, the group also acquired a rival production site. While organizing the site's activity for 1 year, the engineer learned that the site will close to transfer the production lines to the recently acquired site and it was them who were entrusted with this closure: "go ahead, close the factory". Assuring continuity in production while organizing the transfer of production lines, accompanying the reclassification of the 70 employees at the site being closed, these were the activities that they were responsible for, without knowing about their future following the closure of the site.

We see here an event that directly and suddenly affected S' mission, and which was associated with an unanticipated strategic decision for rationalization on their part, because their initial role was to revitalize the production site before making the decision to close it.

The closure of the production site in Brazil, the transfer of the lines to another site as well as the reclassification of the employees of the closed site, was entirely successful. This testifies to the success of an appropriate response to the event.

Box 2.2. *The closure of a production site*

Finally, the flexibility and adaptability required when dealing with the volatility and speed of the markets and the actors and technologies involved mean that collective improvization is necessary. Far from being able to rely on standard procedures or stabilizing routines, on the contrary the activity very often must deviate from accepted control and organizational procedures to respond rapidly to changes and hazards that make the usual commonplace practices inappropriate. Thus, engineers find themselves more and more frequently in situations of collective improvization.

Situation 3: The inability to respond to a client's demand

E. is a consulting engineer within BC practice. Although a paid employee of the practice, her activity takes place, during tasks, principally at clients' places of work. The succession of tasks means that E. chooses them according to the consulting firm's resourcing constraints. However, the last task has not been based on her technical expertise, but rather on a lack of resources: an unanticipated resignation, the launching of projects mobilizing key actors within the practice.

We see here that a client's demand sometimes requires improvisation to respond by mobilizing an inexperienced engineer. However, in the face of these tasks, E. quickly feels helpless, unable to respond to the requirements of the client alone: "We do not have answers to everything, but the problem is that we pay you to have an answer to everything". Faced with this event, the mobilization on a project for which she does not feel competent, answers will be provided despite everything to satisfy the client.

But, at least from the viewpoint of its main protagonist, these answers did not avoid failure. Indeed, totally helpless due to the difficulties that she faced, the Consultant Engineer was forced to resign, in a painful way.

Box 2.3. *The inability to respond to a client demand*

These three situations that we have just introduced illustrate the event-based character of a contemporary engineer's activities. A factory closure decreed by a strategic direction which "falls" to a process engineer, a breakdown in communication associated with intercultural misunderstanding or "parachuting" of an inexperienced engineer onto a consulting mission, these are three events that affect the activities of engineers "involved" in these situations without having anticipated them, and which translate into an "eventful"[9] contemporary organization.

These three situations also show the very heterogeneous level of the results, from success to failure passing through a stalemate situation. What we wish to question in the following section is linked to the response or to the reactions of engineers when faced with these events. Do these responses, when they exist, reveal innovation? We make this assumption by considering that to innovate is to exploit these dynamics that are distinct to a contemporary organization. Thus, the production of new values gained from the use associated with technical and/or organizational devices was inspired by the uncertainty that is generated by this dynamic. As destabilizing as events such as the ones we have described may be, they offer a potential for renewal and evolution. It is in response to these events that we must look for new sources of innovation. Access to innovation means understanding the events and the range of responses incurred.

2.2. Innovating: a transaction or inventive response?

2.2.1. *Exploiting events: two possible interpretations*

We therefore propose to analyze innovation starting from the concept of an event, thus understood as according to the following definition (with A. Badiou and P. Zarifian): the event designates an unpredictable and inanticipatable rupture of a situation, a rupture for which the consequences are complex and wide ranging and affect those involved, without them having control over the situation.

---

9 We will not consider here the events as appearing recently in the activity of engineers. It is first an intensification of the occurrence of these events and second to consider the factors such as the new modes of governance, the opening in network of the productive logic and the place of the client in the process, like new "sources" of events, which affect the autonomy and the control of the engineers in the conduct of their activity.

In this way, the concept of an event makes it possible to shift the focus away from the contemporary company, no longer characterizing it as an organizing device, but more as an entity perturbed by unpredictable events that may exceed its organizational potential.

It is important to distinguish two moments of this event-based dynamic, which we call the moment of the event strictly speaking, its unpredictable upsurge, and the moment of innovation that the event calls for or requires. However, this (post)event dynamic, which is complex, can be thought of in two ways, according to whether we prefer to reflect on the situation and the experience or on a thought about the movement, movement that exceeds the subjects that it "involves". These two interpretations of the event thus engage two figures of distinct but complementary thought: one of situation and the other of movement. The first prioritizes the situated actions or the interactions during a situation that prevail over the actors themselves, without the situation being a place or a spatially determined entity: it rather covers everything that is engaged in the open transactional[10] relations between an organism/organization and its environment. The second prioritizes the disrupting challenge, which is non-localized, which exceeds the situations and the actors.

The duality of the event – as problematized by Louis Quere – comes, in our opinion, from this double interpretation or possible problematization of the event [QUE 06]. This duality also allows us to consider, in an ambivalent way, the radicality of the event, that is to say how complex and wide ranging its consequences are, and the way in which it exceeds the subjects to whom it "happens to" (figure of the movement), maintaining it in the registration of experience and action (figure of the situation). We will thus find this ambivalence in the problematization of the two moments of the event-based dynamic.

2.2.2. *The moment of the event: experience and/or disruptive challenge*

The moment of the event is that of the rupture, the unpredictable irruption and the discontinuity that it introduces. To seize this moment raises a certain

10 Dewey and Bentley propose substituting the concept of "transaction" for "interaction", emphasizing that the individual or group is in a continuous transformation relationship with their environment [DEW 49].

amount of ambiguity. Indeed, as radical as it might be, the event cannot be understood directly or externally, but indirectly, by the actors that it affects. It thus reaches the status of an event, becoming an event only because it "happens to someone". This "someone" can be understood in an extended sense, as an individual or as a group of individuals, human or non-human [QUE 06]. Even when the event exceeds the actors and limits their possibility for description, it is via this proven dimension of the event that the possibility opens up. The two figures of thought, approaches, retain different elements or possibilities for description regarding the feedback on the relationship that the actors[11] have with the event.

The figure of the movement, the most radical one, sees the event as an irruption from the outside, a rupture of the existing situation and of its fields of possibilities, and which means that it overruns and exceeds the actors. This irreducible rupture or discontinuity versus the situation generates a disruptive challenge facing the event. This radicality annihilates any understanding of the event, from the situation, by the actors that it affects. This challenge is translated first, in terms of behavior demonstration, by negativity, given the impossibility of action and the passivity that it tends to generate. Because the event is endured, it causes suffering, pain, distress or even collapse.

Situation 3: The fall of a consulting engineer

The limits with which the consulting engineer – unable to meet the technical requirements of her client – express the pain generated by this event illustrate this negative dimension, a distressing one, that will retain the figure of the movement. "You realize that you amount to nothing in life", "to be completely under water, it is depressing" are the harrowing feelings she expressed and which have led her to resign. Seeing that I was sinking, my family circle said to me, let it go, thus giving up on the other opportunities linked to this project.

Box 2.4. *The collapse of a consulting engineer*

From the viewpoint of movement, the figure of the situation seems to attenuate the radicality of the event. If rupture and discontinuity remain, the events' externality versus the situation is less obvious. More precisely, while unpredictable, the event could "downscale" the situation as it is not yet

11 These two figures are as opposed to a third, the figure of substance that favors actors and their intrinsic attributes. For a more precise distinction between these figures of thought, situation and movement, see [CHO 13]

determined but conditioned by the situation that it disturbs [MEA 32]. Thus, offering the possibility for actors to restore a certain level of continuity in a retrospective manner unveils certain causalities that give the event its situated character. While it leaves the actors deprived, at the core of a disruptive challenge in the figure of the movement, he remains in the figure of the situation, compatible with a logic of action, understood as being at least partially controlled, normalizable and socializable. The relationship of the actors to the event is then less destructive, and provokes rather a disturbance, destabilization or questioning.

An attenuating speech about the unpredictability of the event

Our three example situations converge on the narrative given by the actors *a posteriori*.

In each case, the degree of unpredictability is attenuated, the warning signs are convened, or the explanatory factors of the appearance are identified. In situation 1, on the difficult collaboration between partners for the development of the CRM tool, the intercultural factor is quickly identified as one of the causes why this collaboration suddenly became inoperative. And in the end, the engineer admitted that these intercultural difficulties could have been anticipated. Likewise, retrospectively, the decision to close a production site (situation 2) following the purchase of another is not improbable from a strategic viewpoint; the engineer integrated it rapidly once the decision had been taken. In the end, in the third situation (impossible response to technical requirements of a client), the consulting engineer remembers her fears and the alarm signals addressed to her colleagues, reassuring nonetheless, on the potential difficulties to assume her role in such a project.

Box 2.5. *An attenuating speech on the unpredictability of the event*

In a certain way, the actors in confrontation with the event agree rapidly on a certain intelligibility. This does not prejudge its radicality or of its unpredictability in principle for us, from the viewpoint of the actors to which this event will occur or the response to be supplied.

2.2.3. *The moment of innovation: renegotiating the problems and/or discovering the community*

Irrespective of the radicality, the destabilizing or distressing character of this moment of the event, it requires an answer and generates postevent consequences. This response is also problematized differently according to whether we convene the figures of the situation or the movement and whether it becomes a moment of innovation by the novelty it carries.

The responses provided in the three situations

In the first situation, the answer to get out of this lack of cooperation is mainly through the recomposition of the team involved in the project, both on the French side and the Indian side. Thus, the active participation of a superior made it possible to go beyond the "gender conflicts" in establishing a more hierarchical relationship. The engineer who was initially very isolated also requested the help of several colleagues: "I needed help from them". On the Indian side, a "little hand", according to the engineer, became the favored intermediary for the French team.

In the second situation, the success of the factory's closure, the response was first of all the recruitment dynamic, collective involvement and dealing with this challenge initially faced alone: "you close the factory". On the other hand, the quality engineer responded by changing the roles within the factory and recruiting new personnel during this transition period, which may seem counterintuitive in such a context.

In the third situation, the failure can be perceived as a kind of sedimentation of the situation, "I was hindered", and the event did not produce a response strictly speaking. The alerts, this time unambiguous from the engineer remained ineffective: "do not worry, this will do". No new interaction came to light with the client, who stayed away: "the client understands nothing, it does not matter". The only lever, which is insufficient, is the extension of the working areas by accessing "training modules on a digital platform, in the evening and at the weekend", to get back to the level required. Without making a valued judgement, we were amazed to see no help being provided to the engineer, no additional time negotiated for the project or no revision of objectives.

Box 2.6. *The responses provided in the three situations*

Many response elements indicate a transactional[12] dynamic within the situation and can explain the success, even mitigated, considering the event. Renegotiating the project's composition without modifying its purpose, making changes from the point of view of different stakeholders to collectively achieve the closure of the site and reclassification of the

12 The notion of a "transactional dynamic" here refers to the concept of "transaction" defined by J. Dewey and A. Bentley [DEW 49, p.137] and mentioned above.

employees are examples of reconstructing problems within the situation while its boundary remains relatively stable. If we consider the least successful outcomes or failures, we could also assert a deficit of transactional dynamics. This is particularly pronounced in the failure to respond when faced with a client's technical requirements: the event leads any stakeholders involved to problematize the situation again and to renegotiate it, and to have a common purpose. The engineer remains isolated, left alone to face the event. We could on this basis also invoke a certain transactional deficit in the situation of international collaboration. Certainly, the postevent reality has been renegotiated, rebuilt by different actors. But perhaps it could have been more densely so: the event exposes intercultural conflicts, but also opposes the choice of technical solution, which have only lately been renegotiated (at the project relaunch). We therefore see it through the prism of the situation, responding to the event via the engagement of a transactional dynamic within the situation allows us to exceed the disruptions that it induces. On the contrary, a lack of negotiation alters the quality of the response to the event.

From a situational perspective, the event and the response to this event, although destabilizing for them, remain accessible to the professionals concerned. The event is certainly a source of change and transformation, but remains subject to the primacy of the action, anchored in experience. The event thus becomes the term of a transaction and a source of new intelligibility in the dynamics of the situation. It initiates a transactional dynamic, a logic of investigation within which the subject and the event mutually reconstitute themselves [DEW 93]. The potential action concerning the event certainly remains limited, notably because it has taken place and it cannot be modified, but it opens up possibilities and thus an investigation process will allow us to progressively remove the indetermination from the situation. The response to the event, from a situational perspective, is therefore inextricable from that of the transaction or the investigation, i.e. of different terms within the situation. Confronting the events, and responding to the disruption caused, leads us back to negotiating the integration of the event in a situation.

This response by the investigation allows us to specify and to characterize the innovation that results in the response to the event. In this situated approach, the innovation is due to the potential emergence of new problems and a new collective construction of these problems, without necessarily placing them in a new situation. Innovation would then be comparable to a collective learning dynamic in response to the disruption caused by an event. We therefore suggest that these answers produce incremental innovation, without a radical break from what already existed in the situation.

In an analysis from a movement perspective, we especially retain that it is the emergence of new collectives, irreducible to the previous situation, which prevails. International collaboration mobilizes new contributors until then not previously involved in the project. The successful closure of the site involves a redefinition of roles, recruitment and creation, all of these being completely new interactions. On the contrary, the failure of the consulting engineer will be read as an inability to (re)create a collective to confront the event. Also, if international collaboration offers a result that is so far disappointing, is it not attributable to the mobilization of a new collective that does not fundamentally renew the previous one? This creation of a new group of actors, according to size, has the capacity to invalidate the previous situations, to initiate new situations, assuming the rupture introduced by the event. Thus, by renewing the actors and their roles, the engineer succeeds in leaving a demotivating situation (closing a production site) to create a new recoverable one (successful career change).

The movement perspective prioritizes the excessive character of the event, without erasing the request for a response from the subject faced with the event. Given this chaos, this postevent movement, how can we envisage a response from the subjects, which cannot proceed from a controlled action resulting from experience, while assuming the disruption that the event provokes? This response will possibly imply new arrangements and unprecedented combinations of heterogeneous elements [DEL 77], which inaugurate new situations and a new experimental world. We will then consider the response to an event as being the order for the inventive response, which deconstructed the situation as it had unfolded before the event. Where the response to the event from a situational perspective is an endogeneization of the event within the history of the situation, the response by invention which occurs if we take movement as our analytical approach assumes that the response is played out and is invented outside the situation, without anticipating the consequences.

This response leads us to think of innovation in an alternative way to that exposed previously. Innovation comes from a response to an event. The latter does not necessarily generate innovation but generates conditions or a potential for innovation. Thus, the responses to an event are potentially innovative. To innovate, when using movement, an inventive response to the challenge could thus be generated, as a result of the difficulties encountered.

Innovation involves the construction of new collectives that actively breakaway from the previous situation. To innovate from the event perspective would involve producing a collective to construct problems and unprecedented situations. In other words, since innovation is conditioned by the constitution of new collectives, it may be more radical than when it takes place through negotiation and transactions, probably implying that it is a more gradual process.

At the end of this analysis, it seems to us indispensable to specify a limit. Indeed, innovation can only be considered in relation to its diffuse social character. However, this diffusion process was relatively inaccessible for us during our observations, principally because of the methodological difficulties involved in long-term monitoring. Thus, to attest and more rigorously confirm the innovation status of these responses to an event, it is necessary to seize the implications of these responses for collective practices, their impact and their diffusion within the organization.

2.3. Conclusion of the event to conditions of its innovative exploitation

Table 2.1 summarizes the implications of the event according to whether one favors a figure of thought of the situation or movement.

	Figure of situation	Figure of movement
Subjective impact at the moment of the event	Experience Destabilization	Disruptive challenge Disorientation
Registration of the response	Undergoing situation Transaction	Inventive response layout
Innovation potential	Incremental	Rupturing

Table 2.1. *Implications of the event*

This summary of the elements presented illustrates the different radicality of the event and its consequences, which are expressed according to the figures of thought. If these two figures are associating the event with a dynamic and an instability, the manifestations of this dynamic can be problematized in a very ambivalent way. It is this ambivalence that perfectly translates the expression of “negative experience” by which Quéré presents the event. By applying these two terms, he shows the tension between a thought of the event anchored in experience, referring primarily to human action “in touch” with the event, whereas the second term refers to the radicality of the event that exceeds the actors and their possibility to act, placing them face to face with a challenge to overcome, which does not call for a controlled response.

This ambivalence has major consequences on the potential innovation that may emerge, in a “postevent” dynamic, like a response required to the event. Thus, the type of response – potentially innovative – susceptible to being provided seems inextricable from the radicality with which the event is lived or perceived. In the classification of the action of the situation, we can ask to what extent this register does not attenuate the radicality of the event-based rupture. And in the registering of the disruptive challenge of movement, we can wonder in what measure this register does not exacerbate the radicality to the point that it gives the innovation an almost “miracle” character. Faced with this new form of innovation, inseparable from an event, the question of the responses registered by the innovative engineer remains largely open, in particular regarding the “influence” the actors can have on this dynamic.

We observe therefore that the response ingredients to an event are multiple and can be “drawn” in the sense of collective negotiation or the invention of the collective according to whether we favor a reactive or situated thought. The contribution of this comparison of situations by the figures of thought is therefore not so much a conclusion on a possible falsification or denial of these different problematizations than in the key points that they highlight.

By way of opening, it seems important to us, if we admit that this idea of innovation possibly initiates a response to the event, to question the facilitating conditions of this innovation as a response to the event. We have observed, without determining the radicality of such an event, that the latter can give rise, or not, to more or less innovative responses. Indeed, the potential for innovation does not seem exclusively linked to the radicality of the event and its disruptive consequences. If this were the case, we would be entitled to expect a minor “collapse” toward a technical difficulty for a client and an easier response than facing the announcement of the closure of a production site and the social issues that it raises. It is also not a question of summoning the individual dispositions or aptitudes, which belong to another figure of thought, essentialist or substantialist, contradictory to what we have presented. It seems to us more pertinent to question the facilitator (a potentially converging concept of the two figures of thought), likely to stimulate an adequate response to the event. One condition seems to be the renewal in the potential for action in the face of an event, condition of underestimation in the previous situation, and therefore of an innovative response. This question of the facilitating environment becomes crucial when we consider that innovation is irreducible to a controlled and planned process, and that at the same time it results in the emergence of the unexpected that must be answered. This shift becomes decisive when we think about the management of innovation in contemporary "events" organization, a form of management that does not consist of an activity management or a lack of intervention, but rather prepares the organizational environments susceptible to facilitating the detection and the "correct" responses to events. Displacement implicitly raises training problems and the awareness to this new source of innovation: the event.

2.4. References

[ALT 93] ALTER N., "Innovation et organisation: deux légitimités en concurrence", *Revue française de sociologie*, vol. 34, no. 2, pp. 175–197, 1993.

[BAD 97] BADIOU A., *Saint Paul. La fondation de l'universalisme*, PUF, Paris, 1997.

[CHE 03] CHESBROUGH H., *Open Innovation: The New Imperative for Creating and Profiting from Technology*, Harvard Business School Press, Cambridge, 2003.

[CHO 13] CHOPLIN H., *L'ingénieur contemporain, le philosophe et le scientifique*, Les Belles Lettres, Paris, 2013.

[CHR 97] CHRISTENSEN C., *The Innovator's Dilemma*, Harvard Business School Press, Cambridge, 1997.

[DEL 77] DELEUZE G., PARNET C., *Dialogues*, Flammarion, Paris, 1977.

[DEL 91] DELEUZE G., GUATTARI F., *Qu'est-ce que la philosophie?*, Éditions de Minuit, Paris, 1991.

[DEW 49] DEWEY J., BENTLEY A.F., *Knowing and the Known*, Beacon Press, Boston, 1949.

[DEW 93] DEWEY J., *Logique. La Théorie de l'enquête (1938)*, presented and translated by G. DELEDALLE, PUF, Paris, 1993.

[DRU 85] DRUCKER P., *Les entrepreneurs*, Hachette Pluriel, Paris, 1985.

[HUE 12] HUET F., CHOPLIN H., "L'économie de fonctionnalité comme économie de "coopéraction" : le cas du développement de logiciels", *Projectics*, no. 11, pp. 111–122, 2012.

[HUE 17] HUET F., "Problématiser le capitalisme cognitif", *Les Cahiers Costech*, no. 1, 2017, available at: http://www.costech.utc.fr/IMG/pdf/huet_introduire_et_problematiser_le_capitalisme_cognitif_costech_26sept2016.pdf.

[LE 06] LE MASSON P., WEIL B., HATCHUEL A., *Les processus d'innovation*, Hermes-Lavoisier, Paris, 2006.

[MEA 32] MEAD G.H., *The Philosophy of the Present*, Open Court Publishing Company, Chicago, 1932.

[MOA 09] MOATI P., "Cette crise est aussi une crise du modèle de consommation", *Les Temps modernes*, no. 655, pp. 145–169, 2009.

[MOU 08] MOULIER BOUTANG Y., *Le capitalisme cognitif ou la nouvelle grande transformation*, Éditions Amsterdam, Paris, 2008.

[QUE 06] QUERE L., "Entre fait et sens, la dualité de l'évènement", *Réseaux*, no. 139, pp. 183–218, 2006.

[SEG 12] SEGRESTIN B., HATCHUEL A., *Refonder l'entreprise, La Républiques des idées,* Le Seuil, Paris, 2012.

[WEI 10] WEINSTEIN O., *Pouvoir, finance et connaissance*, La Découverte, Paris, 2010.

3

Innovation within Companies: Changes and Impacts on Our Student Engineer Training Models

3.1. Introduction

The objective of this chapter is to characterize the innovative changes within companies to provide ideas on the required adaptation of training practices in engineering schools. One of the first characteristics of the concept of innovation is its transversality in research: sociology work intersects with those of education sciences, creating an echo within research into management sciences or even economic sciences. These disciplines have innovative approaches that are sometimes different and sometimes contradictory, notably concerning the fields of education and business. Innovation is thus a multidimensional and multidisciplinary phenomenon that occupies a "vast space" that we cannot represent in a consensual way. We have chosen to position innovation at the interaction between different elements at a social, societal, organizational, methodological, technical and managerial level. Our singular approach, as seen previously with innovation design, makes it possible to question the impacts of the evolution of these conceptions of innovation on training and the engineering profession.

From a competitive advantage, innovation has become a necessity, an imperative for companies in order to ensure their sustainable development.

Chapter written by Christiane GILLET and Klara KÖVESI.

Thus, in order to define conceptions of innovation, we recall two approaches: its aims and its scope in terms of change. Technological innovation is the most well known, whereas innovation can take an infinite number of forms other than that of a new process or a radically different product. This can be a new design, a new organization or even new information or a communication system, which can be formal or informal. Likewise, the degree of the innovation varies.

These plural dimensions of innovation and associated processes enable questioning of the organizational forms developed within companies in order to succeed at breathing life into creative ideas: how do we think about the development of innovation today within companies? Also, how is it organized, what are the new practices? Problems connected to the innovation dynamics raise many questions such as fit-for-purpose technology, work methods, organizational culture, R&D activity and management style. The multidisciplinary approach present in the innovation process is noticeably more frequent and simultaneously these dynamics can be more or less formalized within companies.

This chapter is organized into two parts. The first part allows us to specify conceptions of innovation in companies in terms of purposes and the organization of work. The second part makes the link with the stakes that these changes represent for engineering training in terms of theoretical contributions and practical applications. What are the impacts on engineer training following current conceptions of innovation and its structuring within companies? New ways of working exist in companies to meet the challenges of innovation. In what way should our training models evolve so that our engineering students are able to fully participate in the development of innovation integrated into company strategy? Our objective is to envisage how conceptions of innovation in a company can meet those for engineering training in terms of expected skillsets.

3.2. The transformation of innovation within contemporary companies

As Blondel [BLO 10] invites us to, we must discuss the meaning and purpose of innovation. Innovation for what? Also for whom? The misuse of a concept leads it to lose its meaning and can thereby return to negative content assimilations. The injunction to innovate is indeed so present today in discourses – innovate or perish! – that we can also ask ourselves if this is not a myth, an illusion, a syndrome of current times. What exactly is meant by innovation? In classical economic analysis [CAR 83], innovation is very often reduced to its technical dimension and is considered as being a source of technical progress, which is itself considered as being the principal factor for economic growth. However, only 20% of innovations are technical and 80% are social, organizational, commercial, marketing or financial [GOD 10]. Alongside this, the reasons for innovation are also multiple, because innovations always exist within a social environment. They must respond to a given social need in a mature context. It is the users, by their very practices, which give shape and depth to the proposition of progress [GAG 11]. Innovations must be contextualized: any project can result in innovation if it has a market ready to accept change and a solvent demand, just as it can fail if the market does not adhere to it. In order to complete our analysis of contemporary conceptions of innovation, the second development of this first part focuses on new incurred work organizations. We will thus be able to develop, explained in the second part of this chapter, questions related to the required adaptation of training practices in engineering schools.

3.2.1. *Nature and purpose of innovation*

Joseph Schumpeter, an Austrian economist, showed from 1934 in his work theory of economic development [SCH 34] that the nature of innovation – the objects on which it can impact – is a multiform notion. It is important to distinguish and to analyze its sources and consequences: the introduction of new products, the introduction of new production methods, the opening of new markets, the development of new sources of procurement with regard to raw materials or other inputs and the creation of new market structures within a branch of an activity. This classification makes it possible to distinguish invention – the discovery of new scientific knowledge – from innovation. Indeed, concepts such as creativity and invention are often assimilated with that of innovation because they are associated with novelty. But as detailed by Mercier-Laurent, "an invention has little economic value if it is not transformed into innovation addressing a market" [MER 11, p. 48]. Innovation requires risk taking, especially for the marketing of the invention.

Product innovation can involve offering a completely new product or service, whether in its configuration or in its use, for example the television when it was launched after the Second World War. But novelty can also exist independently from use, for example, when the compact disc replaced the vinyl disc. Similarly, product innovation can simply be based on the improvement of an existing product, for example the color television substituting a black and white set. The aim of process innovation is to improve the company's production or distribution process in order to reduce cost and/or improve the quality of relations with suppliers. There is no interface with the client in this type of innovation, but research concentrates on innovation in terms of production methods (including procurement) and distribution.

Information management systems, techniques and equipment are at the heart of thinking on innovation conducted jointly between the research and development service and the production service. The adoption of new forms of work organization in the company – such as Taylorism (division of work) or Fordism (conveyor-belt system or assembly line) – is also an innovation. Organizational innovation involves "introducing a new method in the organization of the processes, activities, work space and the relations between the units and stakeholders in order to improve employee performance" [MER 11]. It is therefore necessary to think differently about the way in which we organize ourselves. The objectives can be to improve company performance in terms of production or distribution methods (process innovation) but may also involve improving the employee satisfaction level at work, or even the possibility of bringing external knowledge into a company. For example, the company's strategy integrates customers and partners in its learning network. We are interested here in the organization of skills and knowledge. We mean collaborative organizational innovation or even coinnovation. Selling a product with the associated services is a marketing innovation, for example, in the automobile industry: a car offer linked to a care and maintenance contract and an offer for a bank loan and insurance. Current trends include a commercial offer regrouping personal services simultaneously, such as shopping, accommodating a child in a crèche or looking after an elderly person. These services are not innovative, because they existed well before the concept of "personal services", but creating a company to combine them or offer them to other companies is a marketing innovation: the way in which products or services are offered to clients is innovative. There is permeability and a synergy between the four forms of innovation presented. A marketing innovation always has an impact at an organizational level and therefore repercussions at the process level. Similarly, product innovation requires improvements in the production process.

To continue our exploration of the forms of innovation and thus to understand the skills required for its development, we are now interested in the degree of change that they create on the basis of two opposing models. First of all, the so-called "radical" or "disruptive" innovation corresponds to the proposal for new technology, a new product or service. This type of innovation creates new markets and new uses. Disruptive innovations do not necessarily imply new technologies. They are more likely to be combinations of existing and proven technologies but brought together in a new product architecture offering the user (individuals, companies) a collection of new attributes. These technologies frequently provide a value proposition that is very different from what the user knew and was used to using. The name "disruptive", which is also employed in this type of innovation, insists on the disruptive aspect of this form of innovation. Disruptive innovations lead to the design of a product with more limited performances, at least in the short term. For Devalan [DEV 06], its realization is difficult because its development is expensive and risky: the demand is not clearly identified; the methods and tools internally are redeployed for an uncertain objective; the skills and resources internally are displaced in a zone of uncertainty. Schumpeter characterized radical innovation as a process of "creative destruction". Innovation destroys value (decline or disappearance of obsolete activities) as it simultaneously creates new sources of wealth. The second model known as incremental innovation involves improving or creating other variants to the products, services or processes which exist already: hybrid car, sustainable packing, improvement of a service. For Rahmouni and Yildizoglu [RAH 11, p. 6], "these innovations are often produced by the companies that do very little in-house research and which do not resort very often to patents and external licenses". The objective is to create and penetrate other markets by having adapted radical innovation to the specifics of the demand. Incremental innovation is similar to the concept of sustaining innovation introduced by Christensen [CHR 11, CHR 13]. For a company, it involves improving a given technology in order to satisfy its clients, to increase its margins and to increase its part of the market facing its competitors. Continuity innovations are the most frequent within companies. They tend to improve the performance of an existing product or a given technology on the basis of performance criteria that average users have already evaluated regarding a market.

For Christensen, disruptive and sustaining dynamics act simultaneously. Also, over the mid and long term, an innovation considered as "disruptive" evolves and the sustaining movement leads an approach that exceeds the existing technology. According to the author, investing massively in disruptive technologies is not a rational decision from a financial point of view. First, because these technologies are generally marketed in emerging markets and are often small in size. Thus, product users do not usually want these new "disruptive" products, because they cannot utilize them in the initial stages because of their weak performance. An important result from Christensen's work is that radical innovation is the form of innovation that allows companies to be successful, whereas numerous case studies tend to show that companies focus on incremental innovations (for routine financial and organizational reasons because the risks taken are less, etc.). Christensen's work and the cases that he documents are mostly about technological innovations in products or production processes. Nevertheless, its analytical framework also applies to various domains, particularly services.

The question of the degree of change reveals that innovation is in fact located in a continuum. Sociologically, innovation has all the traits of a product or a service, which is imposed rapidly and has been crowned with success. For Gaglio [GAG 11], it is in fact the outcome of a process with multiple branches for which it is the provisional result. Just like the concept of crisis, we can compare innovation to an iceberg: the emerged part refers to the visible result; and the immersed part is much more interesting and corresponds to the reality, because innovation is the result of small steps, confrontations internally and externally, preparations, second thoughts and failures. Who are the actors working in this submerged part? Where do they come from? How are they organized and which are the transmission belts which will permit the collective to transform cooperation into a dynamic of innovation? In an underlying manner, the questions are asked on the training of these actors and the role of the human and social sciences involved. What are the key skills that we need to emphasize so that our student engineers are able to meet the challenge of new organizations so that innovation in companies can be achieved?

3.2.2. *The organization of innovation*

A company's innovation process is based on a very sensitive dynamic in terms of management. New management methods emerge and require the replacement of mechanistic thinking by systemic thinking [MOR 99]. Innovation implies the creation of a collective dynamic (innovative spirit) in an innovative company. Segrestin and Hatchuel [SEG 12, p. 45] in their work *Refonder l'entreprise* justify thus the concept of a "modern company": "the great error would be therefore to believe that the invention of the modern company naturally comes from the history of commerce or that it is a simple historic adaptation of market economic logic. The company is breaking with the economic doctrines of its time, to give meaning to innovative collective projects". Innovation also needs to have a collective work space put in place (collaborative environment) dedicated and organized for this activity. All in all, the leadership method must adapt to these changes through a new form of management authority, which is based on professional legitimacy and in turn based on skills (innovative management style).

3.2.2.1. *The innovation process*

The innovation process is often characterized as a way of obtaining added value. For Boly [BOL 04], this added value can appear at several levels: strategic (for example a competitive advantage), functional (a technological advantage), commercial (for example an increase in the share of the market), financial (for example an increase in profit), hedonistic (such as the pleasure experienced by the actors participating in the creation), intellectual (for example new knowledge) and marketing (for example the image or the notoriety of the product). Furthermore, we can name this process "snow ball", because the added value created at a level will have an innovating impact on another sector of the company. It is also a recursive process because the actual innovation reinforces the possibility for future innovation. The current evolution of the innovation process is the sign of a continually growing complexity: "the complexity based on the principle of recursivity, recursive causality: the identified processes are recursive, i.e. only the products from the process generate the process that produced them" [TER 00, p. 13]. Because of these recursive relations, the organization and the innovation process are continuously modified by a coevolution that provides a dynamic for company innovation.

The temporary nature of the innovation process has also become a very significant constraint in the strategy developed by companies and in the progress of the innovation process. Time is a decisive dimension for innovation because it must be organized as fast as possible. The lifecycle of products and services becomes in effect shorter and shorter. This implies shortening the innovation process. To ensure this acceleration and this innovative dynamism, companies must invest more and more in this activity in the short term but have a vision in the longer term. Thus, to take a leading position on the market, the company must have an innovation process, which is faster than the competition.

Often, the innovation process deviates from its initially planned trajectory [GAG 11]. The innovation process is in effect closely linked to the unknown and to uncertainty. The notion of the unknown is a neutral notion in principal. The unknown can be experienced positively as a sort of challenge, discovery or accomplishment of the dream that can bring many benefits. On the contrary, this unknown and this uncertainty can be experienced negatively with the identification of these notions to risks taken when we think outside the box! In France, managing these uncertainties creates difficulties, because "the unknown scares us, and often scares us a lot. Looking for great security, at any price, imposes great prudence on us. But what often causes us problems is not prudence, it is mostly excessive prudence, excess of control, excess in the search for security" [COR 06, p. 296]. This cultural habit of fear of risk taking leads to a slowdown in the dynamism of the innovation process. Yet, risks are considered as an integral part of innovation. In this situation, teaching of risk management is very important at an organizational level.

Organization of the innovation process can be more or less formalized depending on the nature of the process (for example the existence of loops and complexity of technology). In order to have good management of the different steps of the innovation process, a very large number of organizations adopted the Stage-Gate model proposed by Cooper [COO 94]. It is the systematic examination of each innovation project in its different development phases that facilitates the decision to continue or stop the project. The main advantage that explains the popularity of this model is that it allows rigorous management of the innovation process as well as risk reduction with its accurate assessments. Yet, this model which applies a sequential approach for innovation leaves questions concerning its actual effectiveness for certain types of innovation. Because of its significant formalization, it is not very flexible for projects requiring rapid decision making: to seize an opportunity or take a segment of the market before the competitors [ROY 02]. This decision-making model is increasingly being replaced by a sequencing approach, which is designated as a "competing approach to innovation" [LE 11]. In this case, the application of a sequence of conditional decisions allows us to move to the following step of the process, temporarily, pending the result from the decisional evaluation of the current step. Obviously, this approach requires much greater investment in exchange for saving time in carrying out the innovation process. The response is that it represents higher risk taking, because this operating mode implies complex decisions based on multiple constraints and conditions. These complex decisions require experienced project management, coupled with good technical knowledge. The sequential approach of the innovation process is recommended mostly for radical innovations, with the aim of reducing risks. The competitor approach is recommended for incremental innovations, because it allows progress alongside the different steps, making the process more dynamic and more rapid (gaining time).

3.2.2.2. *Collective innovation: configuration between actors*

Communities play an increasingly important role in the innovation process: "they allow improving the company by considering the innovative ideas of their passionate members" [SAR 17, p. 244]. Very often, these communities gather actors from different organizations, because the innovation process requires bringing together knowledge from multiple domains and it implies very close collaboration between numerous jobs and numerous stakeholders coming from different horizons. This implies an extension of employer flexibility and also a change in management method according to the level of the applied technology. The links which unite the community dedicated to innovation and to the company are above all based on trust, acknowledgment and support without expecting anything in return, thus this is goodwill. This enactive approach of the innovation process is based on the actors' representations of their own actions, but also the environment in which they act – notably how they adjust their representations to those of the other actors with which they evolve. Moreover, by the context that the actors give individually and collectively to their actions, they all contribute to shaping the organization's innovation strategy. For Valérie de La Ville and Éléonore Mounoud [LA 06, p. 99], "this 'enactive' perspective, which grants an essential place to the ordinary actions of organization members, enables us to understand how the large diversity of the activities, in which the members of the company are engaged, contributes to the creation of a multiform and ambiguous sphere, providing multiple potential meanings, which have their own internal robustness from the participants' viewpoint".

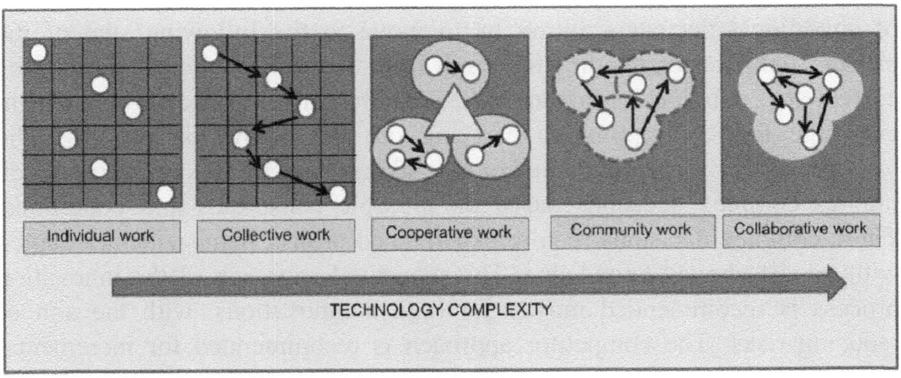

Figure 3.1. *The different forms of work organization at work at the same time*

With the emergence of a team's creative capital [ALB 17], to stimulate and develop the collective intelligence of a community, work organization has had to evolve quickly to respond to the challenges of innovation. The figure above, inspired by the work of Christensen, illustrates the different forms of work organization.

Individual work is found at the first level with a relatively simple technology, that is low level of technicity and complexity. This level creates no organizational problems because it is a simple form of work: the employees working individually by producing work that does not require specific collaboration. At the second level, we find work organized in teams, which is composed of different jobs. In this case, learning about the construction of a veritable multidisciplinary team is very important for collective work. We are in a more complex technological framework, but still inside the same company. In this context, collaborating to perform collective work is much easier, even if the team members belong to different organizational units. At the third level, we find communities of economic interest that implement cooperative work. At this level, we go outside of company's framework by pooling resources and risks with other companies.

In this community, sharing knowledge is not necessary, and most of the time the entities of each participant work autonomously (for example in production cooperatives). At the next level, in the case of the communities of practice, the communication and sharing of knowledge are inevitable. The production of reflexive work requires a rather formal organizational framework by developing communal practices. Virtual communities regroup in virtual environments without a physical existence. They therefore have a rather informal organizational framework. At the higher level, in the case of competing companies we can speak of a veritable collaborative work. At this level, the companies regroup to share their knowledge and thus be able to innovate. This type of work organization enforces the strict definition of a contractual framework for the rules covering the sharing of knowledge, risks, resources, investments and profits. It also requires employee mobility (geographical or professional) between the different external and internal entities to produce common innovation. In this type of work organization, the practicing of nomadism or ubiquity (remote work) is very frequent. To resolve the complexity of intercompany collaborative work, the application of collaborative management is inescapable through putting in place a common platform for running the project [COR 06].

Edgar Morin, at the global congress for complex thought that took place on December 8, 2016, stated: "the word complexus means 'connected', 'woven together' and therefore, complex thinking is a thought which connects, firstly, by contextualizing, i.e. by connecting to context, on the other hand, by trying to understand what a system is". Organized innovation is based on the links that unite the actors participating in its process and its development. They are not based on the hierarchical lines of the organization. At the same time, the innovator, more than being a fantastic inventor, is first and foremost an individual who possesses several strings to his bow, one of which is decisive, that of intra/entrepreneurship – and also that of being a lobbyist, the leader of a network or even the catalyst of ideas. We have put forward some characteristics of the innovation process: complexity, time pressure, uncertainty, risk taking, degree of team work formalization. We have also highlighted that the organization of these processes are based on the multidisciplinary teams often managed from outside the company's framework, and for which the values are based on transparency and respect. The objective is to develop collective intelligence. These thoughts on the new ways of working for engineers and the issues that these evolutions represent for their training are developed in the second part of this chapter.

3.3. The impact of the new forms of innovation design on the training of engineers

Today, for organizations, innovation appears as a question of survival: in a context of intensive innovation [HAT 02], it has a central function in the life of organizations. Its development process has seen significant evolution in the last two decades, with the appearance of new forms of innovation that have repercussions on work practices. It is clear that the engineer has a paramount role to play in the development and management of innovation.

From a managerial point of view, collective management of innovation requires a particularly high level of technicity and complexity to bring together the knowledge and skills of multiple domains, and to converge the work of miscellaneous jobs [COR 06]. Thanks to their technical training which is the core of their job, engineers in general have an adequate level of technicity, but they often have gaps when undertaking collective management for which they are much less prepared for during their initial training. The main challenge of engineering training, from a managerial point of view, is to prepare young engineers to fill this managerial function and thus be capable to establish collaborative management in overcoming the difficulties linked to the complexity in practicing collaborative innovation.

From a functional point of view, the engineer plays an important role in the acceptance of innovation by potential users. Beyond the technical progress and the technological innovation – which have been linked historically to the engineering professions [TER 84] – it is nowadays not only supposed to ensure the technical viability of innovation, but also its societal flexibility. For this, the engineer needs an enlarged vision: they must be capable of technical control in their domain, but also be able to contextualize knowledge to integrate it in a societal context [FLI 03].

In this section, we will discuss the current challenges for engineering training in innovation, issues which are closely linked to the recent transition of the conception of innovation within organizations. What are the consequences of these changes on engineering training? How can we adapt the training to this new operation of organizations? What knowledge and skills should be developed? How can we integrate them in the engineering training curriculum? We will tackle these questions with a critical eye by discussing not only the expectations of the organizations, but also those of student engineers.

From an organizational perspective, we have distinguished two categories of stakes according to their roles in the innovation process. Initially, we will speak of the stakes that are linked to the management of innovation, more precisely the collective management of innovation. Then, we will tackle the issues that have a functional impact on the progress of the innovation process, allowing engineers to complete their mission by the contextualization of their knowledge and its integration in a societal context.

3.3.1. *Managerial aspect of the innovation process*

Collaborative work is considered as the foundation of collaborative innovation. According to Christofol *et al.*'s definition [COR 06, p. 71], "collaborative work is collective innovation in a contractual framework of sharing knowledge, risks, costs and profits". According to this definition, collaborative management mostly concerns the engineers' capacity to carry out relational management, knowledge management and risk management between the members of the collaborative team.

3.3.1.1. *Relational management*

Since collaborative working teams are composed of very miscellaneous participants and are often numerous (for example employees, experts, intercompany work teams, communities of practices, virtual communities or networks of companies), reaching a satisfactory level of collaboration requires efficient collective management. Therefore, relational management occupies a particularly important place in the collective management of innovation. Following recent transformations within organizations concerning the practice of innovation, we find ourselves in a situation where the teams must unite to design everything, in transdisciplinary collaboration. In this context, according to IESF recommendations (society for the engineers and scientists in France), "the engineer permanently performs their role in complex ecosystems, vast webs of interdependencies which are intensely collaborative, where the principal engine is the objective to cooperate, and the most important skill is relational capacity: understanding others, making yourself understood, knowing how to listen, knowing how to express and make yourself heard, admitting that others can have different and nevertheless pertinent points of view, knowing that to admit to others that our point of view is also legitimate, knowing how to negotiate, knowing how to jointly prepare for shared purposes and satisfactory solutions" [IES 16, p. 9]. Even if today the development of this relational capacity for young engineers is not displayed among the priorities for engineering training, it will probably take an ever more important place in the future, for the simple reason that it has a direct influence on the performance of collective innovation. Today, relational management is supposed to consider not only the diversity of collaborative work teams, but also the mobility of the participants. The ever increasing propagation of collaborative work in a virtual environment has added a new dimension to relational management: being capable of managing international virtual teams with the help of technology. The remote management of a virtual team requires engineers to apply a very different management style and assumes an organization of work that is very different from proximity management [KAR 10].

3.3.1.2. *Knowledge management*

Innovation is closely linked to knowledge, because it unites "the creation and the application of new knowledge to make it productive" [FOU 06, p. 169], via an approach to knowledge capitalization that appears as a major challenge of intensive innovation generated by high competition on the market. In the context of a knowledge-based economy, contemporary organizations impose the development of efficient knowledge management to meet the ongoing necessity for new knowledge [HAT 02]. Today, collaborative learning, which concerns the capacity of a team to learn collectively and produce collective intelligence, has become the dominant strategy for organizations to obtain new knowledge. It is a collective training experience that collides with miscellaneous obstacles (cognitive, relational, organizational) during its realization [KÖV 15]. For engineers, implementing efficient collective learning management involves not only being capable of overcoming these obstacles, but also having the capacity to bring together knowledge from multiple and varied domains. Engineers must also be capable of operating in an extended organization for interorganizational collaborations. This requires a capacity to organize the sharing of individual skills and to "synergize resources coming from multiple actors to create new skills" [DEF 13, p. 47]. These new managerial practices, often operating in an open innovation framework [CHE 17] by removing the border between organizational knowledge, must face many obstacles linked to the sharing paradox [BOG 11] and the protection of knowledge during collaborative work [ATT 15], for example the legal management of intellectual property for a new product or the contractualization for knowledge sharing during interorganizational collaborative work. Even if these problems appear to be far from its domain, an engineer is considered as being capable of detecting them, taking them into consideration and analyzing them, and thus managing them despite a framework that is not always or not sufficiently formalized. With respect to knowledge transmission within organizations, engineers traditionally play a role in the transmission of technical knowledge. With the new practices for innovation which come clearly from technical domains, engineers have the obligation not only to transmit technical knowledge or other explicit knowledge (for example concepts, structures and documents), but also tacit knowledge (such as expertise, an experience and a memory). This also requires engineers to exchange between these two modes of knowledge [BAL 02].

3.3.1.3. *Risk management*

Controlling risks linked to collaborative innovation by identifying "potential events susceptible to affecting the organization and managing the limits of its risk appetite" is considered by contemporary organizations as a major strategic issue [MAN 09, p. 18]. A particularly important managerial function for risk management is its contribution to decision making by identifying the potential risks and by proposing an action plan with particular attention paid to strategic activities of the organization. In the risk management process, the most difficult is without doubt the identification of risks, above all in the case of radical innovation without previous experience. In this case, engineers are required to call on their explicit knowledge such as intuitiveness, subjectivity or imagination, to pass from one knowledge domain to another in the knowledge cycle (chaos → complex → knowable → known) [SNO 07] in order to be able to propose a pertinent action plan.

As contemporary organizations evolve in a more and more uncertain world that changes perpetually, there are many risks linked with the direct external environment of organizations (see Figure 3.1). The four reactions possible to risks are as follows: avoidance, reduction, sharing and acceptance, which innovative organizations apply according to traditional models to avoid and reduce risks [MAN 09]. For organizations developing new forms of collective innovation, the sharing of risks between partners is a frequently strategic choice. However, this practice of sharing requires interorganizational management of risks with a high level of complexity and uncertainty.

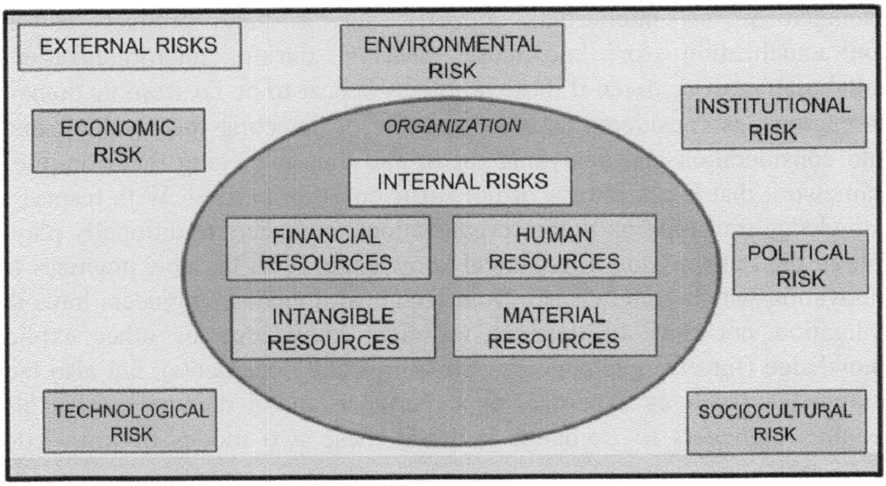

Figure 3.2. *Risk categories in organizations*

Risks linked with the internal environment of organizations concern mostly available resources (human, physical, financial and intangible) [BLO 06]. We emphasize, in particular, the importance of human resources that condition the success of innovation. For example, the loss of skills and disengagement of the employees are frequent situations in organizational life, which endanger the success of innovation. Engineers' skills, more particularly the skills linked to their explicit knowledge (technical knowledge, concepts, etc.), with a high level of specializations, represent a significant risk for organizations.

3.3.2. *The functional aspect of the innovation process*

In collective innovation practices, engineers are present at every step of the innovation process, thanks to the varied activities their job entails: the design phase involves engineers, designers and experts; in the creation phase of new products, there are development engineers and testing engineers; and in the evaluation phase, the commercial engineers contribute by making their knowledge available to the service of innovation. From a functional point of view, they are a precious supply for innovative agencies. They share a global vision, thus facilitating the adoption of innovation, and instill an entrepreneurial spirit to help its evaluation.

3.3.2.1. *The global vision*

We live in a society of growing complexity, in which it becomes essential for engineers to have a global vision allowing them to understand and to see the world more broadly and, in its entirety, to take account of increasingly complex problems. This global vision presupposes openness and the development of "general intelligence" by having extended knowledge. According to Edgar Morin, "the global is more than the context, it is the whole containing diverse parts interoperatively or organizationally related" [MOR 00, p. 44]. According to this definition, the engineers of our time are supposed to have the ability to understand their knowledge in its entirety. Contrary to traditional ideas that defend the superiority of technical knowledge for engineers, "the more powerful the general intelligence, the greater the ability to deal with special problems" [MOR 00, p. 44]. Close to the notion of global vision, the diversity of thought (diversity in thought) unites the capacity of engineers to practice not only convergent thought, but also divergent thought. Divergent thought has been defined by Guilford [GUI 68] as a source of creativity using intuitive intelligence, in leaving conventional frameworks. Divergent thought is considered by contemporary organizations as a potential resource to stimulate innovation. Despite the recognized potential of divergent thought, it is convergent or standard thought that dominates the teaching and training of engineers, because of the sometimes inert character of the system facing changes. This requires first, conventional and organized learning before leaving formalized frameworks behind and practicing divergence.

Even if it is united in the concept of the global vision, we emphasize the importance of contextualization for engineers in innovation action, i.e. thinking about technology with society [FLI 03]. Considering the social context of innovation is essential throughout the innovation process to ensure successful adoption of the new product by the consumers or by the end users. Today, the principal issues linked to innovation training are as follows: first, developing understanding of complexity by global thought so that young engineers are capable of solving very complex problems; second, they are capable of thinking jointly about the technical and the social aspects, and thus applying a sociotechnical approach to innovation. Therefore, it is very important for young engineers to develop multifaceted and well-grounded skills and technical, scientific and human training during their initial training [GIL 16].

3.3.2.2. Entrepreneurial spirit

Today, while organizations have ever more difficulty in following the pace of innovation, they look for engineers with entrepreneurial spirit and who want to practice entrepreneurship (corporate entrepreneurship). Intrapreneurship is a relatively recent concept defined by Chung and Gibbons [CHU 97, p. 14] as an: "organizational process for transforming individual ideas into collective actions through the management of uncertainties". In other words, entrepreneurship concerns the reproduction of an entrepreneur's attitudes and behaviors for the profit of an organization. This requires the engineers to not only have knowledge in technical or management methods, but also to develop a state of mind "through values such as risk assessment, the initiative or the objective and general attitudes such as the responsibility or the objective to change" [FAY 03, p. 38]. The transmission of these values, attitudes and the entrepreneurial culture of the young engineers should occupy an important place in engineering training. Among these values, we will highlight two which appear to us to be particularly linked to the contemporary socioeconomic context: risk awareness and the sense of responsibility.

Following the work of Hofstede *et al.* [HOF 04], the avoidance of risks and uncertain situations is traditionally anchored in French society. Instead, it favors stability and predictability. Risk taking does not appear as a positive act. It is particularly true for the engineering profession, among which only a very low percentage have conducted an entrepreneurial activity [IES 15][1]. In this rather unfavorable cultural context for entrepreneurship, the development of risk awareness should be a priority in engineering training to incite the young to undertake and to innovate: to give them the right to fail and attenuate their fear of a possible failure, which has a negative image within society. In the case of continuous innovation, risk taking is negligible, but it becomes a key element in case of breakthrough innovation. For organizations with an innovation strategy based on the development of disruptive innovations, the predisposition of their employees to take risks is indispensable for their performance [LE 11].

With respect to responsibility, we retain the social character of this value and insist on the link between social entrepreneurship and innovation, because the consideration of the ethical and social characteristics of

1 In 2015, only 3.8% of the engineering population conducted an entrepreneurial activity unlike 10.1% of the active population in France.

innovation is an important phenomenon within contemporary organizations. The social responsibility of the intra/entrepreneur is closely linked with social changes for the improvement of social conditions, globally or locally [PER 06]. Given the importance of this question in today's society, social entrepreneurship has appeared very recently in some engineering training curricula. This remains sporadic and, in most cases, training is integrated in entrepreneurship projects, often upon the request of student engineers [GIL 17].

3.4. Conclusion

Successful collaboration in a process dedicated to innovation assumes a level of skill being possessed by the participating actors. "In a world where only agility and the rapid relationship between contributors will allow innovation, finding new solutions and generating value, cooperation takes precedence over subordination" [COU 14, p. 318]. Reinventing team management to better understand and mobilize actors and to liberate their creativity is a condition for success in the innovation process. The engineering training curriculum must also follow this movement so that students can grow and evolve in these fast moving and agile professional environments. How can we prepare them? By reinforcing in schools these principals for multidisciplinarity in training teams, by multiplying the intercultural projects, between institutions, by soliciting the students on roles and positions previously unheard of, by organizing encounters between manufacturers and students, etc. The objectives are to remove students from their comfort zone, preparing them to work in a context of uncertainty so that they can train themselves, for example, to identify the risks and thus manage them. By giving them the opportunity to evolve in training in unknown but nevertheless welcoming environments, we encourage them to organize autonomously, to take responsibility, to define together their action plan, to take at certain moments the leading role and to define all the values of their working group.

In the innovation process, work practices are disrupted, because we assist in an extension of cooperation. The boundaries of companies disappear and mix. Innovative projects are now transdisciplinary. For example, new social requirements can contribute to developing new technologies. Engineers must be able to examine themselves and therefore be able to know the world. Good ideas and those which receive positive feedback are often the result of collaborative work between the different professions. The future subsidiaries created are proof of this. Connective health, for example, will be rich in innovations over the next decade, generating a complete range of new jobs, at the boundary of medicine and engineering. The development of sensors combined with the ageing population allows certain innovation processes in the medical and paramedical categories to see the light of day. Even if the world of engineers still crosses only very little with that of doctors, it is safe to bet that the health sector and the engineering world will be required to cooperate further. Companies require more and more complex skills to work on these synergies. New jobs and new activities come to light and their success depends on human factors, which are an integral part of a company's capital. Schools and universities must consider these emergences.

3.5. References

[ALB 17] ALBASIO B., CRAVERO G., *Créative attitude*, Dunod, Paris, 2017.

[ATT 15] ATTOUT A., BARBAROUX P., "Nouvelles approches de l'innovation et gestion des connaissances: quelle articulation? Quels enjeux? Quels changements?", *Réseau de Recherche sur l'Innovation*, vol. 14, 2015.

[BAL 02] BALLAY J.F, *Tous managers du savoir*, Éditions d'Organisation, Paris, 2002.

[BLO 06] BLONDEL F., GAULTIER–GAILLARD S., "Comment une entreprise peut-elle maîtriser les risques induits par l'innovation?" *Vie & sciences de l'entreprise,* vol. 172, no. 3, pp. 10–23, 2006.

[BLO 10] BLONDEL D., *Innovation et bien-être: une relation équivoque,* Publibook, Paris, 2010.

[BOG 11] BOGERS M., "The open innovation paradox: knowledge sharing and protection in R&D collaborations", *European Journal of Innovation Management,* vol. 14, no. 1, pp. 93–117, 2011.

[BOL 04] BOLY V., *Ingénierie de l'innovation: Organisation et méthodologies des entreprises innovantes*, Hermes-Lavoisier, Cachan, 2004.

[CAR 83] CARRE J.-J., DUBOIS P., MALINVAUD E., *Abrégé de la croissance française,* Le Seuil, Paris, 1983.

[CHE 17] CHESBROUGH H.-W., APPLEYARD M.-M., "Open innovation and strategy" *California Management Review*, vol. 50, no. 1, 2017.

[CHR 11] CHRISTENSEN C., *The Innovator's Dilemma*, Harper Business, New York, 2011.

[CHR 13] CHRISTENSEN C., RAYNOR M., *The Innovator's Solution. Creating and Sustaining Successful Growth*, Harvard Business Review Press, Boston, Massachusetts, 2013.

[CHU 97] CHUNG L. H., GIBBONS P. T., "Corporate entrepreneurship: The roles of ideology and social capital", *Group & Organization Management*, vol. 22, no. 1, pp. 10–30, 1997.

[COO 94] COOPER R., "Third-generation new product processes", *Journal of Product Innovation Management*, vol. 11, no. 1, pp. 3–14, 1994.

[COR 06] CORSI P., CHRISTOFOL H., RICHIR S., SAMIER H. (eds), *Innovation Engineering: the power of intangible networks*, ISTE Ltd, London and John Wiley & Sons, New York, 2006.

[COU 14] COUTENCEAU C., *L'intelligence économique au service de l'innovation*, Éditions Eyrolles, Paris, 2014.

[DEF 13] DEFELIX C., PICQ T., "De l'entreprise étendue à la gestion des compétences étendue: enjeux et pratiques en pôles de compétitivité" *GRH*, no. 2, pp. 41–66, 2013.

[DEV 06] DEVELAN P., *L'innovation de rupture: clé de la compétitivité*, Hermes-Lavoisier, Cachan, 2006.

[FAY 03] FAYOLLE A., "Instiller l'esprit d'entreprendre dans les grandes entreprises et les organisations", *Gérer et Comprendre*, vol. 72, pp. 26–39, 2003.

[FLI 03] FLICHY P., *L'innovation technique: Récents développements en sciences sociales. Vers une nouvelle théorie de l'innovation*, La Découverte, Paris, 2003.

[FOU 06] FOUCHÉCOUR DE M., "Knowledge Management for Innovation", in CORSI P., CHRISTOFOL H., RICHIR S., SAMIER H. (eds), *Innovation Engineering: the power of intangible networks*, ISTE Ltd, London and John Wiley & Sons, New York, 2006.

[GAG 11] GAGLIO G., *Sociologie de l'innovation*, PUF, Paris, 2011.

[GIL 16] GILLET C., KÖVESI K., COADOUR D., "Adapting engineering training to the new challenges of a fast-changing innovative working environment", *23th Educational Innovation in Economics and Business,* EDHEC Business School, Nice, France, June 8–10, 2016.

[GIL 17] GILLET C., KÖVESI K., "Social entrepreneurship education for engineering students", *24th Annual Conference of Educational Innovationin Economics and Business*, HU University of Applied Sciences Utrecht, The Netherlands, May 31–June 2, 2017.

[GOD 10] GODET M., MOUSLI M., DURANCE P., *Libérez l'innovation dans les territoires,* La Documentation française, Paris, 2010.

[GUI 68] GUILFORD, J.P., *Intelligence, Creativity and their Educational Implications*, Robert R. Knapp, San Diego, 1968.

[HAT 02] HATCHUEL A., LE MASSON P., WEIL B., "De la gestion des connaissances aux organisations orientées conception", *Revue Internationale des Sciences Sociales*, no. 1, pp. 29–42, 2002.

[HOF 04] HOFSTEDE G., NOORDERHAVEN N. G., THURIK A.R. *et al.*, "Culture's role in entrepreneurship: self-employment out of dissatisfaction, Innovation, entrepreneurship and culture. The interaction between technology, progress and economic growth", in BROWN T.E., ULIJNE J. (eds), *Innovation, Entrepreneurship and Culture*, Edward Elgar Publishing, Cheltenham, 2004.

[IES 15] IESF Société des ingénieurs et scientifiques de France, 26e enquête nationale sur les ingénieurs, survey, 2015, available at: http://home.iesf.fr/offres/file_inline_src/752/752_P_36601_577e6330ce9bf_8.pdf.

[IES 16] IESF Société des ingénieurs et scientifiques de France, Repenser la formation des ingénieurs pour répondre au défi de la complexité, survey, 2016, available at: http://home.iesf.fr/offres/doc_inline_src/752/D4_Former_les_ingenieurs_a_la_complexite.pdf.

[KAR 10] KARJALAINEN H., SOPARNOT R., "Gérer des équipes virtuelles internationales: une question de proximité et de technologies", *Gestion*, vol. 35, no. 2, pp. 10–20, 2010.

[KÖV 15] KÖVESI K., "L'apprentissage du travail collaboratif dans un environnement virtuel: pour mieux comprendre les nouvelles pratiques collaboratives", in CARDONA GIL E., LEMAÎTRE D. (eds), *La modélisation des activités managériales au défi de la formation, Analyse d'un seriousgame*, L'Harmattan, Paris, 2015.

[LA 06] LA VILLE V.-I. (DE), MOUNOUD E., "La créativitéde l'agir ordinaire – éléments pour une approche stratégique", in GOLSORKHI D. (ed.), *La fabrique de la stratégie: une perspective multidimensionnelle*, Vuibert, Paris, pp. 91–107, 2006.

[LE 11] LE NAGARD-ASSAYAG E., MANCEAU D., *Marketing de l'innovation de la création au lancement de nouveaux produits*, Dunod, Paris, 2011.

[MAN 09] MANDZILA E., ZEGHAL D., "Management des risques de l'entreprise: Ne prenez pas le risque de ne pas le faire !", *La Revue des Sciences de Gestion*, vol. 3, pp. 237–238, 2009.

[MER 11] MERCIER-LAURENT E., *Innovation Ecosystems*, ISTE Ltd, London and John Wiley & Sons, New York, 2011.

[MOR 99] MORIN E., LE MOIGNE J.-L., *L'intelligence de la complexité*, L'Harmattan, Paris, 1999.

[MOR 00] MORIN E., *Les sept savoirs nécessaires à l'éducation future*, Le Seuil, Paris, 2000.

[PER 06] PERRINI F., VURRO C., "Social entrepreneurship: Innovation and social change across theory and practice", in MAIR J., ROBINSON J., HOCKERTS K. (eds), *Social entrepreneurship*, Palgrave MacMillan, New York, 2006.

[RAH 11] RAHMOUNI M., YILDIZOGLU M., Motivations et déterminants de l'innovation technologique: Un survol des théories modernes, Working document, université d'Aix-Marseille II and III, 2011.

[ROY 02] ROYER L., "Les procédures décisionnelles et le développement des nouveaux produits", *Revue française de gestion*, vol. 28, no. 139, pp. 7–25, 2002.

[SAR 17] SARAZIN B., COHENDET P., SIMON L., *Les communautés d'innovation*, Éditions EMS, Paris, 2017.

[SCH 34] SCHUMPETER J.A., *The theory of Economic Development*, Harvard University Press, Cambridge, 1934.

[SEG 12] SEGRESTIN B., HATCHUEL A., *Refonder l'entreprise*, Le Seuil, Paris, 2012.

[SNO 07] SNOWDEN D.J., BOONE M.E., "A leader's framework for decision making", *Harvard Business Review*, vol. 85, no. 11, pp. 68–77, 2007.

[SOL 56] SOLOW R.M., "*A contribution to the theory of economic growth*", *Quarterly Journal of Economics*, vol. 70, no. 1, pp. 65–94, 1956.

[TER 00] TERMI L., "Les processus d'innovation: une approche par la complexité", *IXe conférence Internationale sur le Management Stratégique*, Montpellier, France, May 24–26, 2000.

[TER 84] TERNIER A., "Être ingénieur d'hier à aujourd'hui", *Culture Technique*, no. 12, 1984, available at: documents.irevues.inist.fr.

4

Skills and Competencies for Innovators: New Priorities and Requirements for Engineering Graduates

4.1. Introduction

In our knowledge-based society, the central role of skills and knowledge for innovation is undeniable in industrial business organizations. The term "innovation", however, covers a wide range of activities and requires serious investments that are commonly viewed as a key requirement for value creation. For Grant [GRA 96, p. 112], "the critical input in production and primary source of value is knowledge". Applying the knowledge-based theory of value, "all human productivity is knowledge dependent, and machines are simply embodiments of knowledge". Therefore, organizations' human resources are considered as the main source of innovation. From this perspective, the most efficient way to prevail in intense global competition is to develop a strong human capital and exploit the core competencies defined as the collective skills and knowledge of organization.

Given the ongoing acceleration of the product development cycle, to attain sustainable competitive advantages and high performances, organizations have to be able not only to follow market changes but they should be always one step ahead of others. With the words of Porter [POR 90, p. 75] "innovation and change are inextricably tied together. But change is an unnatural act, particularly in successful companies; powerful

Chapter written by Klara KÖVESI and Péter CSIZMADIA.

forces are at work to avoid and defeat it". That is to say innovation in any case requires "investment in skills and knowledge, as well as in physical assets and brans reputations". Innovation, however, cannot be restricted to new product development. In our understanding, innovation is a value creation process that leads to the development of new ideas and concepts into new products, services or organizational solutions.

It is obvious that there is a general trend toward an increasing need of skills and competencies in today's emerging knowledge-based economy. In a competence-based competition [HEE 97], the lack of needed skills and competencies seriously reduce innovation capability and the economic performance of organizations. This is particularly true in the industrial sector where the development of innovative products and services is closely related to engineers' skills and competencies to innovate. As a result, for industrial business organizations the skills and competencies of their engineers have a strong effect on their capacity to innovate. Which kind of skills and competencies are needed exactly for engineers to enhance innovation?

Global competition and technological advances have made profound changes on the labor market requiring continuously updated knowledge and competencies for engineers with altering priorities. These changes are noticeably perceptible in the industrial sector with the extension of required skills and competencies for graduate engineers enable them to innovate. To enhance their employability[1] potential, they have to be able to use new skills like social and emotional intelligence, adaptive capacity, sense-making, fast learning, convergent and divergent thinking or virtual collaboration, etc. What is the perception of industrial business organizations? Do newly graduated engineers really have these new social and functional skills and competencies to fulfill their requirements?

In this chapter, we propose to explore these new priorities and requirements for engineering graduates' skills and competencies to enhance innovation and highlight the importance attached to them in the industrial sector. Our main purpose is to examine this competency gap between the expected performance level by industrial companies and the perceived performance that recently graduated engineers deliver. In the first part of this

1 We define employability as the ability of engineering students to find and keep a job.

chapter, we propose a state of the art review of literature on the emergence of industry requirements for engineers' skills and competencies. In the second part, based on the results of our qualitative research study, we provide empirical evidence about the industry perception of freshly graduated engineers' skills and competencies to innovate. Our aim is to create a better understanding of the coherence of the companies' innovation skill requirements and the set of competencies young engineers possess in a contemporary Hungarian context. The other reason for restricting the sample to multinationals was the aim to understand to what extent global challenges and initiatives are influenced by the local context. As the target of the investigation is both description and analytical interpretation, qualitative research methodology was chosen in order to explore and understand the heterogeneous views of the various stakeholders.

4.2. Which skills and competencies are needed for innovation?

Generated by the rapidly increasing technical advances, there is an ongoing tendency with a constant extension of future engineering work skills and competencies. Still, this extension concerns predominantly not the technical knowledge but the addition of new non-technical skills and competencies. In this section, we propose a literature review to build a theoretical background of skills and competencies needed for engineers to innovate with an overview of emerging tendencies in industrial organization.

4.2.1. *Toward a holistic vision of engineers*

For a long time, engineering skills and competencies were reduced to technical skills and knowledge. Traditionally, technical skills and knowledge embody the core competencies of engineers such as sciences and engineering fundamentals, engineering applications, engineering practice, etc. In the context of industrial society, the main requirement of industry for engineers was only having strong technical skills and knowledge. At this time, engineers were concerned with the technical aspect of their profession. Consequently, engineering education was historically concentrated on the development of these core technical skills and knowledge until the early 1980s when the economic and industrial environment fundamentally changed.

As a result of emerging technological advances, a profound restructuration has occurred in the traditional industrial sector from manufacturing industries toward service industries. With globalization, the competition was extended to the international level and became more and more intensive in the postindustrial society. On account of these profound changes, engineering practices and the role of engineers have meaningfully changed. This transition engendered a gradual adjustment of their skills and competencies to this new global competitive environment. From this time on, "engineers must supplement technical mastery with business and communication skills, and an understanding of the ethical and societal impact of engineering solutions" [LAN 99, p. 43].

There is a general long-term tendency toward a continuous extension of engineering skills and competencies that concern essentially non-technical skills and competencies and only to a low extent technical knowledge. This ongoing tendency of the extension of engineering competencies is clearly observable in several national engineering competency classifications[2]. It is beyond debate that for engineering education technical knowledge remains the core competency with a predominant place in the engineering curriculum. However, the proportion between technical and non-technical subjects in the curriculum has significantly changed with an increasing fraction of the last one.

This growing importance of non-technical skills and competencies for engineers is explained on the one hand by the increasing complexity of sciences and technology as a result of technological advances and diversity. To develop and successfully commercialize a new innovation, organizations need a vast variety of highly specialized knowledge and competencies. On the other hand, engineers have to take into consideration not only the technical aspects but economic, environmental, social or societal aspects for their innovation activities. Definitely, the role of the engineer has significantly extended including a large variety of new functions such as manager, leader, entrepreneur and negotiator generally related to non-technical skills and competencies.

2 UK-SPEC Standard for Professional Engineering Competence in the UK, Engineering Competency Model created by the AAES in the USA or The List of Professional Competencies for Engineering Profession edited by the CNISF in France.

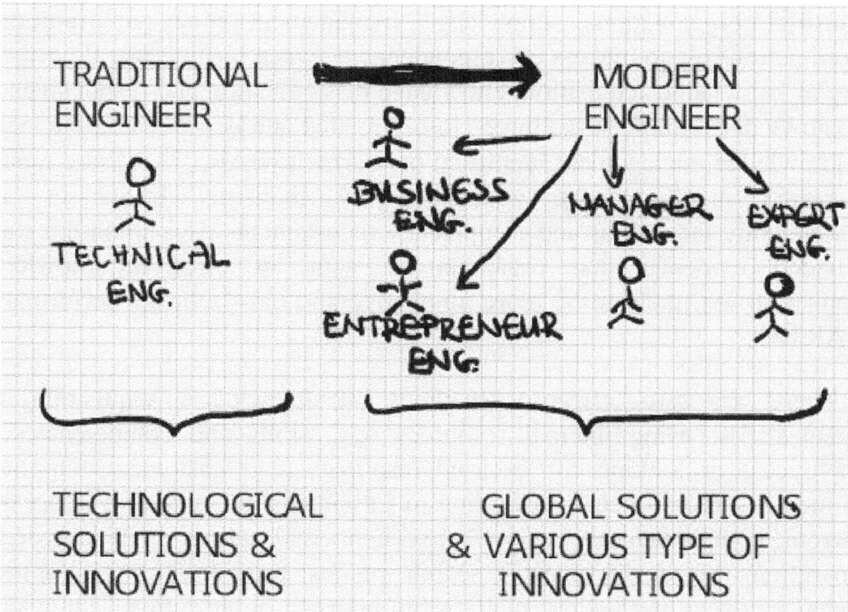

Figure 4.1. *Traditional versus modern engineers' innovation activity*

In this context, to achieve a successful innovation, engineers are required not only to work on a technical solution limited to their field but also to work together with other disciplines to find a comprehensive solution.

As we mentioned before, the engineering profession is traditionally associated with technical innovation based on their solid scientific and technical knowledge. With this extension of the role of engineers, their innovation activity has also been extended. It concerns not only technical but other forms of innovation such as organizational, marketing and social innovations. In section 4.2.2, we discuss industries' changing requirements for engineers' skills and competencies in accordance with their innovation strategy.

4.2.2. *Emergence of industry requirements*

As several recent research works suggest [FOR 16, ALE 13, LOK 09], organizations' skills and competencies have a direct and undeniable influence on their innovation capacity and performances. Consequently, industrial organizations' knowledge management plays an important coordination role to make a correspondence between resources and needs by way of knowledge acquisition, knowledge dissemination and responsiveness to knowledge. Amongst the three closely related and dependent components of knowledge management identified by Darroch [DAR 03], we are mainly interested in knowledge acquisition by industrial businesses and consequently their skills and competency requirements. Her empirical study provided evidence that organizations with an effective knowledge acquisition are likely to be more innovative and have a better performance [DAR 05].

There are two effective methods of knowledge acquisition for organizations: using internal resources (e.g. skills and competencies of employees) or looking for external resources (e.g. customers, suppliers, competitors, recruitment). According to Cassiman and Veugelers [CAS 06], a strong complementarity exists between internal and external knowledge generation. Moreover, their research results provide strong empirical evidence concerning a close relationship between organizations' innovation activities and innovation strategy.

First, we are focusing on the generation of external human resources, namely the generation of individual skills and competencies for innovation by recruiting freshly graduated engineers. Second, we explore the relationship between organizations' innovation strategy and their skills and competency requirements for engineers. Finally, we discuss the impact of the innovation process on organizations' skills and competency requirements.

4.2.2.1. *Effective knowledge generation*

For adequate knowledge generation, organizations should correctly identify their requirements in order to satisfy their needs in terms of skills and competencies for innovation. They should also correctly define what kind of skills and competencies, technical or/and non-technical, are needed for their engineers to foster innovation. They should then outline their priorities between these skills and competencies.

The identification of organizations' requirements and priorities between technical and non-technical skills and competencies is closely related to their market environment. For Song *et al.* [SON 05], there is a complementarity between technical and non-technical skills and competencies that create a positive synergy for innovation. However, this synergy is highly variable depending on the market environment of industrial organizations and "managers and researchers frequently fail to take into account the moderation effects of environmental contexts" [SON 05, p. 270].

Based on Mintzberg's classification[3] of environment [MIN 79], the environments of industrial business organizations differ depending on the competition level in an industry (hostility), the stage of the lifecycle of industry (hostility, dynamism), the dominating technology change (dynamism), the interconnectivity between environmental elements (complexity), the number of stakeholders that organizations deal with (diversity) and the resource availability (hostility). Between the four elements of classification, dynamism has been revealed to have the most significant effect on innovation [PER 14].

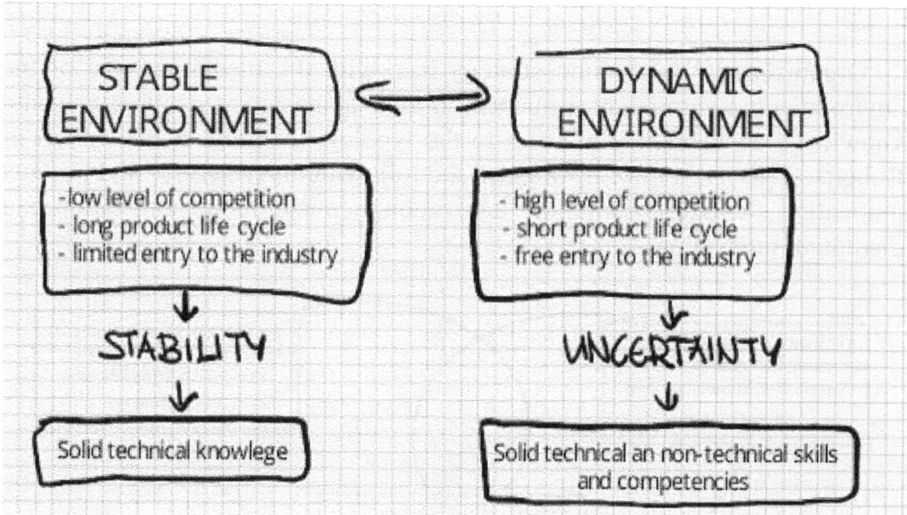

Figure 4.2. *Link between organizations' environment and skills and competency requirements*

Accordingly, in a stable environment with a low level of competition and long product lifecycle, organizations' skills and competency requirements are based on ongoing activities. In this case, organizations should prioritize

3 The four components of Mintzberg environment classification are dynamism, complexity, diversity and hostility.

solid technical knowledge and competencies of their engineers for the sustainability of their innovation actions. For these organizations, the technical skills and competencies of their engineers appear as a human resources asset that enables them to keep their place in the market by limiting the possibility to new entry and limiting the competition.

On the contrary, in a dynamic environment, industrial business organizations perceive a hard competition in their industrial sector with very short product lifecycles dominated by rapid technology changes. This dynamism is intensified due to the higher uncertainty that arises when new players may freely enter an industrial market. In this case, complementing engineers' solid technical knowledge with non-technical skills and competencies becomes a priority to ensure better adaptation to this rapidly changing environment.

4.2.2.2. *Influence of product innovation strategy*

Organizations' innovation strategy has, with the mediating role of the environment, a significant influence on their competency requirements and knowledge acquisition. Referring to the typology of product innovation of Robertson [ROB 71], we distinguish the following three service and product innovation strategies: continuity innovation, semicontinuity innovation and disruptive innovation.

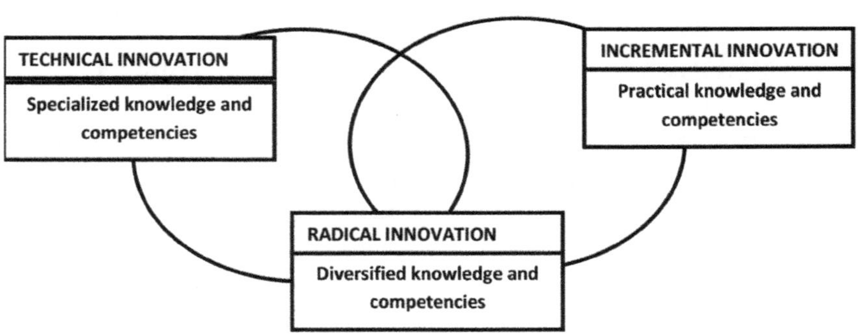

Figure 4.3. *Knowledge and competencies for different types of innovation*

In the case of continuity innovations that are generally related to incremental product innovations with a low level of technological changes, industrial organizations need engineers who understands and are able to manage the ongoing technological solutions. Tushman and Anderson [TUS 86] described these innovations as "competence enhancing innovations" as they require skills and competencies concerning old technology for developing new innovations and products. For this reason, engineers' practical knowledge and competencies became a priority in the application of this strategy. In this situation, it is not surprising that organizations opt for strengthening the technical competencies of their engineers.

For semicontinuity innovations with a high level of technological or usability changes (the way the final consumer uses it) but keeping the same functionality for consumers, organizations need engineers who don't only have a strong technical background and good product knowledge. Having a very high level of specialized knowledge and competencies allows them to develop new functions for existing products or to substitute an existing product due to new technological solutions[4]. As in this case the development of these specialized skills and competencies are required for engineers, these kinds of innovations are also considered as competence-enhancing innovations with the reinforcement of technical and non-technical skills and competencies. This reinforcement should be put into practice by building on existing skills and competencies and extending or complementing them.

Finally, we consider the adoption of disruptive or radical innovations, "which incorporate a large degree of new knowledge organizational complexity (the number of different occupational specialties) and the depth of the organization's knowledge resources (the number of technical or engineering personnel), organizations' need a diversified knowledge". For Dewar and Dutton [DEW 86, p. 1424], the diversity of specialized knowledge is essential for radical innovation as "the more different types of knowledge that are present, i.e., the more complex or specialized the organization, the higher the rate of radical innovation adoption". To develop a new product with a very high level of novelty, these kinds of innovations are qualified as "competencies destroying innovations" that require a fundamentally new set of skills and competencies.

In conclusion, we would like to highlight the fact that in a rapidly changing environment the continual development of skills and competencies is included in the practice of engineering professions that require a high

4 Examples: mechanical → electric typewriters, LSI → VSLI semiconductors.

capacity of knowledge adaptation and continuous lifelong learning. Consequently, one of the most important abilities for engineering graduates is their capacity to learn and use new skills and competencies.

4.2.2.3. *Influence of process innovation strategy*

The traditional view describes innovation as a technology- and/or science-based linear process moving from the invention through design and production to the market (final consumer). In this approach, innovation is a unique and isolated event and is dominantly restricted to R&D efforts. Innovation, e.g. creation of something novel, is a complex phenomenon encompassed by a conceptual fuzziness. Recent theoretical approaches, however, provide a different perspective emphasizing the "recursive" character of the innovation process. This means that the innovation process can be triggered by various causes and is accompanied by complex feedback mechanisms where interaction between various actors and institutions and continuous learning play a crucial role [FRE 87, LUN 88, EDQ 97, SCH 01].

Lundvall [LUN 07] elaborated a dichotomy model of innovation differentiating the basic approaches. The Science-Technology-Innovation (STI) mode of innovation produces novelty on the basis of intensive use of scientific and technical knowledge and high R&D expenditures. In the STI mode analytical, codified and explicit knowledge plays a decisive role. Innovators often cooperate with academic knowledge producer institutions. This mode is typical in high-tech industries, e.g. among pharmaceutical, biotechnology, nanotechnology, etc. companies. On the other hand, the second approach is based on practice-based learning by Doing, by Using and by Interacting (DUI). In the case of the DUI mode of innovation, knowledge is mainly practical, tacit and highly contextualized, and the necessary learning takes place on the job, as innovators face permanent changes and are pressed to create new problem-solving mechanisms. The output of the innovation process is developed by exploiting and utilizing implicit knowledge and learning takes place in collective, intra- or interorganizational networks based on intensive interactions with different stakeholders (employees, suppliers, consumers, etc.). Synthetic knowledge based on technical experience is of utmost importance here; not surprisingly this mode of innovation is typical in engineering-based industries, such as the machinery and automotive industries [JEN 07, LUN 07, PAR 16].

The main difference between the two modes of innovation is that STI is science based, whilst DUI is experience based [LUN 04] and it requires different types of interactions and learning accordingly. In the case of the STI mode, the main challenge is how to absorb scientific knowledge, and in the case of the DUI mode the central issue is the ability to exploit implicit (tacit) knowledge. Applying the STI mode of innovation, emphasis should be put on the scientific and technical knowledge of engineers. Consequently, industrial organizations require strong basic scientific knowledge, especially in engineering sciences (e.g. mechanics, electronics, etc.). On the contrary, if applying the DUI mode of innovation, they attach more importance to engineers' non-technical skills and competencies and prioritize the capacity of engineers to apply their scientific knowledge.

However, the question remains still open as to what skills and competencies are needed at the individual (employee) level to support both modes of innovation? Innovation skills are the broad range of knowledge, skills and competencies that make individuals capable of thinking and acting innovatively in the field in which they are working and operating. These competencies are mainly a combination of technical (project management, IT-skills, etc.) and non-technical (communication, problem solving, etc.) skills. Based on the results of our empirical research study, in section 4.3, we present a typology of skills and competencies for young engineers expected by industries to increase innovation.

4.3. Industry perception of graduate engineering students

As innovation has taken a central role in industrial organization activities, the mastering of innovation skills and competencies has become vital for young engineers. But what kind of skills and competencies are needed to fulfill industrial organizational requirements? Do graduate engineering students really have the skills and competencies to innovate? What is the perception of industrial organization?

We have conducted a qualitative survey to collect empirical evidences about skills and competencies to fulfill industrial organizations requirements to innovate [KOV 16]. We have carried out 15 semistructured interviews with different industrial organizations employing a proportionally high number of engineers in engineering-based industries, operating in Hungary. Our sample was chosen in the view of ensuring the better heterogeneity of these industrial organizations as they were selected from various industries and different regions of Hungary. Also, employees at different levels from top management to freshly employed young graduate engineers were interviewed. As large, multinational companies are at the forefront of innovation in the Hungarian economy, the selection was restricted to them.

Based on the finding of our literature review and the results of two pilot interviews, we developed the final structure of the interview guide composing the following three topics: work experiences with new graduate engineers, their perceived competencies, innovation activities of young graduate engineers. Then, we accomplished a thematic analysis as a practical and flexible methodology to have a comprehensive understanding of the research subject. We identified common issues and categories in our textual database, divided them into subcategories and investigated the emergence of each pattern in order to achieve the most inclusive understanding. For analytical purposes, we followed the taxonomy presented above and differentiated between technical and non-technical skills and competencies.

4.3.1. *Technical competencies*

We define technical competencies as the aggregation and synergy of scientific knowledge and functional competencies. Scientific knowledge includes a well-established scientific and technical background in engineering sciences (e.g. mechanics, electronics and robotics) considered as the core competencies of the engineering profession. However, it is evident that having scientific knowledge, considered as a "good foundation", is indispensable but not sufficient for the practice engineering profession. It is fundamental for engineers to also have functional competencies considered as their ability to put scientific knowledge into practice, which is strongly required by industrial organizations.

4.3.1.1. *Scientific knowledge*

Having a strong scientific knowledge in technical fields is considered as self-evident for the engineering profession. This is particularly true for new graduated engineers as companies do not have a duty to teach them this knowledge. As was outlined by a senior research manager "the theory must be taken from the university". However, a solid technical background embraces not only factual knowledge but it is also "important to have an engineering mindset". This engineering mindset involves the way of thinking in order "to invent new things, to be open-minded to try new solutions and to be able to learn from our own mistakes". The manifestation of the engineering mindset considered as a main recruitment condition "because if they have it, then it is sure they will deal with the problem and find a solution".

Because of their high level of specialization, industrial organizations are aware that recently graduated engineers' specializations would rarely be suitable and directly applicable to their activity. Consequently, they have to establish an internal "on board training" system to teach them relevant and usually highly personalized specific knowledge well-fitted to their activities. As was pointed out by a senior expert "in the first two-three months they do not know what to do, they only learn. And then they start working at three-six months, alongside someone". In general, these internal trainings aim not only to teach new employees the basic background of a highly specialized technical field but to teach them how to use it in an effective way for their everyday work. With the words of a senior manager, newly employed young engineers learn, "the practical work in informal ways as well, with the help of a contact network during the problem solving processes". Industrial organizations have integrated these specific knowledge transfers in a formal and informal way into their training process. However, this internal training requires an adaptation period as was explained by a top manager "when we plan human resources, we count a person…in the first three months as a zero, second three months as a half, and then after that as a whole".

4.3.1.2. *Functional competencies*

From an operational point of view, functional competencies of graduate young engineers are vital for industrial organizations. Their new employees' ability "to put into practice their theoretical practice…and solve the problem" is amongst the basic employment requirements. These functional competencies have become so important that numerous industrial organizations have included verification practices in their recruitment process (e.g. for electrical engineers specializing in speech coding this could involve the development of a specific coding algorithm). In the recruitment process, they ask the applicant to solve short case studies to find a practical solution to a concrete problem. As a senior manager in a multinational telecommunications company revealed, "…when we see that someone has less knowledge but solved the task, even with our help, it is more positive". In this case, the specialized technical knowledge takes a secondary position; the most important are young graduate engineers' functional competencies, especially their problem-solving capacity. Nevertheless, young graduate engineers have in general only a basic theoretical background without any significant experience in a company. For this reason, they are not in a good position to be hired compared to experienced engineers.

Industrial organizations are aware of "the fact that our system works internally it will obviously not be taught by this university", so they rarely require specific functional competencies related to a specialized technical knowledge. In spite of this, they request a broad system vision and the ability "to think in a holistic way". A senior software engineer illustrated the lack of this system vision through an example where young engineers "have to test for an entire system, but they don't even know the basics of the system". This lack of a holistic vision of technical knowledge could be explained by the existing separation between the various scientific subjects in their engineering curricula. In many cases, they do not have the sufficient opportunity to work on huge projects handling a complex system that demand the active use of knowledge and competencies of different scientific subjects at the same time. Controversially, the broad introduction of interdisciplinary teaching that would be a good solution for this problem is not sufficiently widespread nowadays. As an exception, higher apprenticeship engineering graduates have the opportunity to work on complex projects in a real work situation during their studies. However, as modern engineers have to be able to handle a complex system in its entirety, in a real working environment the separation of these subjects has no meaning.

4.3.2. *Non-technical skills and competencies*

As mentioned before, in the case of innovation, especially when it comes to the DUI mode, the key question is how implicit knowledge can be exploited effectively. The DUI mode of innovation is based on the assumptions that knowledge is embodied in routines and everyday practices held by different stakeholders, therefore, besides technical expertise and experience, interactions and interactive learning are key elements in developing innovative outcomes [LUN 05]. This requires a shift from the "technical to the social" [WAR 06] and those elements of the interpersonal relationship, like communication, empathy, aesthetic appearance and emotional control, attitudes and predispositions, etc., that were conceived as personal characteristics in the past, are becoming more and more important in the innovative working practices. In our research, we devoted special attention to the non-technical aspects, especially to the social skills and competencies. The results can be summarized as follows.

The interviewees' general experience was that the relatively high level of technical expertise of the young graduates is in several cases accompanied by a weakness in soft skills or social competencies. The term "*soft skill*" was often used by the interviewees, its meaning, however, remained unclarified. Despite the conceptual heterogeneity, based on the interview results we have identified the following areas of non-technical (social) skills that are of particular importance in value creation.

4.3.2.1. *Initiative and risk taking*

The most important skill or ability the interviewees perceived as a general shortage or weakness of young graduates is their readiness to work independently accompanied by the lack of initiative. A typical complaint was that most of the junior engineers are not able to work alone and their work requires close supervision, which does not mean a short-coming in their technical knowledge but is rather concerned with their general relationship to risk and responsibility taking. The most often mentioned criticism in this respect was that new entrant engineers often lack a willingness to engage in new work tasks challenging their skills. According to the managers interviewed, young engineers perform their prescribed work more or less precisely and at a high professional quality, but when they finish with the tasks delegated to them, they stop and go no further. "They're not self-sufficient at all. They do not ask questions when they have doubts or even ask the same question five times or more. It indicates to us then, that they have understood nothing".

With regard to that problem, two basic attitudes were identified. People representing the *narrow* or *restricted* attitude approach the labor process as being a linear system of discrete and specialized work tasks, even of high complexity. Those with more *holistic* or *expansive* attitudes, however, treat work as a complex system of interdependent tasks and actions. The recursive or DUI mode of innovation is strongly linked to the second attitude as it provides an opportunity for reflection and learning from mistakes. These, however, require taking more responsibility and risk, autonomous decision making, the ability to leave the personal "comfort zone", to cope with uncertainty and willingness to learn. It is strongly connected to individual capabilities as well as to institutional/organizational factors. The attitude to work depends on the manner of professional socialization, the patterns of behavior transmitted by the educational institutions and such innovative organizational practices that support autonomous work, like a learning-friendly work environment, collective knowledge sharing, adequate incentive system, lack of blame culture, etc.

The problem around attitudes toward risk taking in the labor process has other consequences, as well. Both the theoretical literature and empirical evidence suggest that autonomy in work is associated with an increased innovation capacity both at the individual and organizational level [CED 12]. The willing rejection of autonomy and responsibility taking we observed in the case of young engineers may act as a serious barrier in enhancing creativity and boosting innovation.

Our findings challenge the argument that innovative work organizational practices (like lean production, HPWS, project-based work, etc.) are the most important prerequisites of increasing firms' innovation capacities not exclusively in procedural issues but also in case of development of new products and services [LAM 05, MIN 79, SCH 04, VAL 09]. According to our results, innovative work organization practices do not lead *automatically* to the increase in innovative capacity. We have to stress that they instead serve as important frames and inhibitors, but individual contributions and the development of soft skills (including such personal attributes like attitudes to risk taking) are also needed for exploiting the possibilities they offer.

4.3.2.2. *Cognitive style*

The second group of soft skills is related to the cognitive or thinking style of the young graduates, e.g. the way they think, perceive and process information [RID 91]. The most common shortage mentioned here was the weakness of "engineer style thinking". We identified the following dimensions of young engineers' perceived cognition:

– analytical thinking: this term means a coherent and logical frame of cognition and the ability to reduce complex things into single components;

– structured problem solving: most of our interviewees emphasized as one of the most important attributes of the engineer was the constant effort to identify and solve complex problems. These efforts include critical attitude, sequential problem analysis and the ability to transform problems into alternative possibilities and the systematic evaluation of them. It also requires innovativeness, e.g. producing new ideas based on the analysis of technological processes and providing solutions to improve them. It is based on the effective and systematic utilization of all available information, especially the achievements and failures of prior work;

– systematic and holistic perspective: contrary to the analytical skills presented above, a holistic approach requires the synchronous and system-level interpretation of problems, structures and facts and the recognition of the interconnectedness of various patterns and objects that form larger systems. Planning skills also belong here.

These cognitive-type aspects of engineers highlight the importance of an often neglected dimension of innovation, namely the "learning by experimenting". Experimentation and especially learning from failures connect causes and effects and lead to heuristic understanding of generally hidden connections between various phenomena of a different nature. Risk taking, initiative and the cognitive elements presented before are important prerequisites of this type of learning and innovation. In order to exploit technical experiences, the social dimension of non-technical skills should also be taken into account.

4.3.2.3. *Interpersonal skills*

Innovation and learning are interconnected processes [LUN 04]. As referred to earlier, especially in case of the DUI mode of innovation, knowledge is accumulated in organizational routines and experiences and is embodied in relationships between different actors. In the attempts to utilize collectively accumulated (mainly implicit) knowledge, there is a more intensive need for interpersonal rather than individual actions in the workplace. It, however, requires the ability to work and learn collectively and the readiness for "inter-professional communication". Perspective taking, understanding and tolerating the opinion and experiences of others is a must. In this respect, this area is rather problematic in some cases. "Sometimes I have the feeling that they (e.g. young engineers) fight against the world. They always know best and can't take criticism. It makes communication and cooperation extremely difficult and often leads to bad solutions".

Since knowledge accumulation and exploitation takes place in social networks, one other important aspect of interpersonal relations is the ability that may be labeled as "cultural awareness and adaptability" and covers the ability to accommodate various situations with culturally different environments and people with different backgrounds. This skill refers to the essential understanding and proper interpretation of codes, rules, customs, habits and perspectives different environments provide and the quick adaptation to changes of them. It also serves as an important prerequisite of project-based work.

Communication is also an important part of interpersonal relationships. In the interviews, the importance of both oral and written communication skills has been emphasized with a light critical attitude toward the perceived erosion of the written communication standards. The most important shortage in this respect was poor presentation skills. Presentation is a complex term with various aspects, but in this context it means to get the message concerned with new ideas and/or work results across to the listeners in an effective and persuasive manner. In this sense, it is strongly similar to sales activities. The next quotation represents the introverted attitude that is perceived to be typical among fresh engineering graduates and acts as a serious barrier to effective presentations. “Have you done the job? – Yes. – Does it work? – Yes. – Please may I ask you to present it to us, then? – What should I tell you? It works”.

4.4. Conclusion

In case of the innovation-driven multinational companies operating in Hungary, it seems that the prominence of technical competencies still serves as a basis of engineering practice. Furthermore, non-technical competencies are becoming central for engineers to being innovative. As we pointed out previously, there is a trend for increasing importance given to non-technical competencies by industries, explained by their expectations of recruiting well-trained young engineers with a high capacity to innovate. Also, we would like to underline that there is a complex and multifaceted interaction between these skills and competencies at the individual and organizational level to indicate the results of our analyses at the micro- and mesolevel. Our empirical results can be extended to other contexts, as well, as global economic and technical changes may represent similar challenges in other countries. Our results confirmed that multinational companies competing on international markets are facing similar problems in terms of efficient exploitation of the engineers' innovation skills. The central issue here is to exceed the technology-oriented attitudes of the young engineers and to support them in the development of their non-technical competencies as well as to better integrate these skills and competencies in the innovation process.

In view of that, it would be beneficial to include technical and non-technical skill and competence development in an integrated way in the engineering curriculum, taking into consideration their interaction. As we mentioned before, there is a growing importance for knowledge diversity for industries to innovate. For encouraging knowledge diversity, it would be interesting to further introduce inter- and multidisciplinarity courses in the engineering curriculum. These courses would make a connection between various disciplines that have traditionally been separate. However, this separation is no longer justified as in a real working environment they are used together in everyday practice of modern engineering profession. In addition, better integration of non-technical skills and competencies into the engineering curriculum would facilitate the transition of graduate engineering students from the educational system to the workplace. It would be valuable to design these courses by recreating similar working conditions to what graduate engineers will have in the real world of work. We consider that the best way to do this is to create more collaboration between universities and industries, accompanied by new, innovative training methods (learning by error, project-based learning, business games, etc.). As we pointed out previously, there is an important gap between industry requirements and their perception of new graduate engineers' skills and competencies to innovate. In the interest of reducing this gap and meeting expectations, the relationship between engineering education and industries should be reconsidered.

With the world of Kamp [KAM 16], "no other profession unleashes a greater spirit of innovation than engineering"[5] and innovation will remain at the center of engineering activities. Future innovators have to face technical, economic, cultural, social and societal challenges in their professional life in tomorrows' engineers world of work. They have to be able to operate in a multidisciplinary, multinational and multicultural working environment and be required to apply a "global engineering" approach. This "global dimension is the sum of the social, political, technological, cultural and environmental issues which are shaping engineering at the global level" [BOU 08, p. 4]. It is completely in agreement with the fact that innovation progressively crosses disciplinary boundaries and requires multi- and interdisciplinary thinking from engineers.

5 In his recent book about the future vision of engineering education, he described our world as full of volatility, uncertainty, complexity and ambiguity[6]. In this VUCA world, as he named it, he identified three converging driving forces: globalization and digitization, the horizontalization of the socioeconomic world and the blending of technical, economic and societal cultures.

4.5. References

[ALE 13] ALEGRE J., SENGUPTA K., LAPIEDRA R., "Knowledge management and innovation performance in a high-tech SMEs industry", *International Small Business Journal*, vol. 31, no. 4, pp. 454–470, 2013.

[BOU 08] BOURN D., NEAL I., *The Global Engineer: Incorporating Global Skills within UK Higher Education of Engineers*, IOE, London, 2008.

[CAS 06] CASSIMAN B., VEUGELERS R., "In search of complementarity in innovation strategy: Internal R&D and external knowledge acquisition", *Management Science*, vol. 52, no. 1, pp. 68–82, 2006.

[CED 12] CEDEFOP, *Learning and Innovation in Enterprises*, Research Paper No. 27, Publications Office of the European Union, Luxembourg, p. 173, 2012, available at: www.cedefop.europa.eu/files/5527_en.pdf.

[DAN 02] DANNEELS E., "The dynamics of product innovation and firm competences", *Strategic Management Journal*, vol. 23, no. 12, pp. 1095–1121, 2002.

[DAR 03] DARROCH J., "Developing a measure of knowledge management behaviours and practices", *Journal of Knowledge Management*, vol. 7, no. 5, pp. 41–54, 2003.

[DAR 05] DARROCH J., "Knowledge management, innovation and firm performance", *Journal of Knowledge Management*, vol. 9, no. 3, pp. 101–115, 2005.

[DEW 86] DEWAR R.D., DUTTON J.E., "The adoption of radical and incremental innovations: an empirical analysis", *Management Science*, vol. 32, no. 11, pp. 1422–1433, 1986.

[EDQ 97] EDQUIST C., *Systems of Innovation: Technologies, Institutions and Organizations*, Pinter /Cassell Academic, London, 1997.

[FOR 16] FORÉS B., CAMISÓN C., "Does incremental and radical innovation performance depend on different types of knowledge accumulation capabilities and organizational size?", *Journal of Business Research*, vol. 69, no. 2, pp. 831–848, 2016.

[FRE 87] FREEMAN C., *Technology Policy and Economic Performance: Lessons from Japan*, Pinter, London, 1987.

[GRA 96] GRANT R.M., "Toward a knowledge based theory of the firm", *Strategic Management Journal*, vol. 17, no. S2, pp. 109–122, 1996.

[HEE 97] HEENE A., SANCHEZ R. (eds), *Competence-Based Strategic Management*, John Wiley & Sons, London, 1997.

[JEN 07] JENSEN M.B., JOHNSON B., LORENZ E. *et al.*, "Forms of knowledge and modes of innovation", *Research Policy*, vol. 36, pp. 680–693, 2007.

[KAM 16] KAMP A., Engineering education in the rapidly changing world. Rethinking the mission and vision on engineering education at TU Delft, TU Delft Faculty of Aerospace Engineering, Delf, 2016.

[KOV 16] KOVESI K., CSIZMADIA P., "Industry perception of new engineering graduates: the gap between requirements and reality", 43rd Annual Conference of the European Society for Engineering Education, Tampere, Finland, September 12–15, 2016.

[LAM 05] LAM A., "Organizational innovation", in FAGERBERG J., MOWERY D., NELSON R. (eds), *Handbook of Innovation*, Oxford University Press, Oxford, 2005.

[LAN 99] LANG J.D., CRUSE S., MCVEY F.D. *et al.*, "Industry expectations of new engineers: A survey to assist curriculum designers", *Journal of Engineering Education*, vol. 88, no. 1, pp. 43–51, 1999.

[LOK 09] LOKSHIN B., VAN GILS A., BAUER E., "Crafting firm competencies to improve innovative performance", *European Management Journal*, vol. 27, no. 3, pp. 187–196, 2009.

[LUN 88] LUNDVALL B.-Å., "Innovation as an interactive process: From user-producer interaction to the National Innovation Systems", in DOSI G., FREEMAN C., NELSON R.R. *et al.*, (eds), *Technology and Economic Theory*, Pinter, London, 1988.

[LUN 05] LUNDVALL B.-Å., "National innovation systems – analytical concept and development tool", *DRUID Tenth Annual Summer Conference*, Copenhagen, Denmark, June 27–29, 2005.

[LUN 04] LUNDVALL B.-Å., "Innovation system research and policy: where it came from and where it might go", *CAS Seminar*, Oslo, Norway, December 4, 2004.

[MIN 79a] MINTZBERG H., *The Structuring of Enterprises: A Synthesis of the Research*, Prentice-Hall, Englewood Cliffs, p. 512, 1979.

[MIN 79b] MINTZBERG H., "An emerging strategy of "direct" research", *Administrative Science Quarterly*, vol. 24, no. 44, pp. 582–589, 1979.

[PAR 16] PARILLI M.D., HERAS H.A., "STI and DUI innovation modes: scientific-technological and context-specific nuances", *Research Policy*, vol. 45, pp. 746–756, 2016.

[PAY 04] PAYNE J., The changing meaning of skill, Oxford, SKOPE Issues Paper 1, 2004, available at: http://www.skope.ox.ac.uk/wordpress/wp-content/uploads/2014/04/Issues-Paper-1.pdf.

[PÉR 14] PÉREZ-LUÑO A., GOPALAKRISHNAN S., CABRERA R.V., "Innovation and performance: the role of environmental dynamism on the success of innovation choices", *IEEE Transactions on Engineering Management*, vol. 61, no. 3, pp. 499–510, 2014.

[POR 90] PORTER M.E., "The competitive advantage of nations", *Harvard Business Review*, vol. 68, no. 2, pp. 73–93, 1990.

[RIN 15] RINGEL M., TAYLOR A., ZABLIT H., "The most innovative companies 2015 – four factors that differentiate leaders", The Boston Consulting Group Report, 2015.

[ROB 71] ROBERTSON T.S., *Innovative Behavior and Communication*, Holt McDougal, Madison, 1971.

[SCH 04] SCHIENSTOCK G., "Organisational innovations: some reflections on the concept", in BAMMÉ A., GETZINGER G., WIESER B. (eds), *Yearbook 2004 of the Institute of Advanced Studies on Science, Technology and Society*, Profil Verlag, Munich/Vienna, pp. 137–180, 2004.

[SON 05] SONG M., DROGE H., CALANTONE R., "Marketing and technology resource complementarity: an analysis of their interaction effect in two environmental contexts", *Strategic Management Journal*, vol. 26, no. 3, pp. 259–276, 2005.

[VAL 09] VALEYRE A., LORENZ E., CARTRON D. *et al.*, Working conditions in the European union: work organization, Report, Publications office of the European Union, Luxembourg, pp. 1–66, 2009.

[WAR 06] WARHURST C., THOMPSON P., "Mapping knowledge in work: proxies or practices?", *Work, Employment & Society*, vol. 20, no. 4, pp. 787–800, 2006.

PART 2

New Skills and Adaptation to Training Systems

5

The Training of Innovators between Skill Acquisition and Construction of an Individual Socioprofessional Identity

5.1. Introduction

Training for innovation, given the immoderate use of these terms, requires upfront clarification, notably for the notion of innovation. This is what we have been able to do with regard to our PhD, which was devoted to this subject during theoretical research and empirical investigation (immersions, interviews, observations, collection and analysis of data from various sources) based on innovation training in the French higher education system. On completion of this study, we were able to establish that the different meanings of the term correspond to the different ways for innovation training and that of innovators. We thus propose in this chapter to present the results of our study for the case of French Engineering Schools.

An initial study of the scoping notes of the *Commission des titres d'ingénieur* (CTI)[1] shows that from 2006 to 2016 the content linked with the term "innovation" has evolved greatly, moving from research to company creation; from the creation of new activities via assuming more responsibility. Innovation training was initially assimilated to the development of creativity, and then more to entrepreneurial activity and the creation of economic values, becoming a mindset and included in uses beyond technology. What we describe here refers to entrepreneurship and

Chapter written by Tiphaine LIU.

1 The CTI is the official accreditation body for French Engineering Schools.

descriptions of innovative entrepreneurs, notably in the English-speaking world [CHR 13, DES 12]. The innovative engineer's approach resembles management sciences: dividing into skill sets what an innovator is to better train for each one of these skills.

What also appears in these documents, in a subtler way because it is presented as a matter of course, is that the structure, organization and culture of engineering schools are considered as being the adequate framework for such training.

However, this evidence can be questioned: the teaching in engineering schools is restricted from one side by the curriculum and themes imposed by the CTI, and from the other side by traditions linked to the schools' reputation, for which former student's associations are influential and fervent guardians. In such a framework, can we transmit an appetite for the unknown and novel? Put differently, does teaching innovation require oneself to be pedagogically innovative? Or can we train innovators according to a traditional teaching framework?

We will first try to specify the notions, which will allow us to differentiate two kinds of innovators that correspond to two teaching paradigms. Then we will study different curricula (masters) for innovation training in order to show their distinct issues. Finally, we will develop the concept of an emancipating environment, which is particularly interesting from the perspective of training innovators.

5.2. What is innovation? Who are the innovators?

Etymologically, innovation is "the introduction of a novelty into a custom, into a use, into an act" (dictionary of the Académie française, 4th edition, p. 934).

In the *Que sais-je?* book on the sociology of innovation, Gaglio [GAG 11] uses the criteria of the sociologist N. Alter to define innovation starting from the fact that "the emergence of new social practices in the wake of novelty (in a market, in an organization). [...] each time, the novelty becomes innovation through uses".

Indeed, innovation is the result of a social process that connects the conception of an idea (individual or collective) to its actual use by users (members of one or several social groups) through its materialization via a product, service or device and the organization of its diffusion, which differentiates it from the invention solely centered on the creation of novelty [ALT 02].

This dimension centered around the notion of "novelty" seems for us to be interesting when questioning the context of reflection and the definition of innovation as a training objective by engineering schools. The question is what novelty are we talking about for innovation?

Should renewal be understood as described by the economist J. Schumpeter [SCH 04]: a cyclical temporal vision in which all creation will generate destruction, but also renewal? It is then a societal vision of growth where companies (as well as countries, economic issues now being the mainstay of politics) must be constantly renewed to remain competitive, survive and perpetuate the cycle. Looking for innovation is thus located in the economic logic of destruction/creation to maintain the current system.

However, there is another sense of novelty and innovation in a perspective that is not cyclical, but linear. The philosopher H. Bergson [BER 13] treats innovation as an impulse coming directly from the unsatisfied nature of humankind. In *L'évolution créatrice* (1907), he develops the idea of "permanent creation of novelty". He describes the vital impulse that drives people to innovate. According to him, there is no plan "already planned" or foreseeable. Evolution is always unpredictable, because the world "invents itself unceasingly", without which the route created behind him does not exist. Bergson indeed talks of "radical novelty". It would then involve "opening up to the possible, by definition completely unpredictable" [MEN 11]. This innovation completely redistributes the cards.

Two possible conceptions of novelty and thus of innovation then emerge:

– cyclical innovation as a force for economic adjustment: via innovation, the economic system can continue its cycles, passing from equilibrium zone to equilibrium zone by means of a crisis and the creative capacity of innovators. Change modeling is then possible;

– radical innovation as a subversive force of social transformation: radical novelty creates a new era that breaks with the previous one and creates a completely unpredictable future.

Indeed, we have been able to find these two poles of representations of innovation in the discourse of the people that we have met and interviewed as part of an empirical study on innovation. Altogether, more than 30 interviews have been conducted, but this work also relies on an ethnographic approach of the innovation environment with multiple sources: informal discussions, participation in thematic events, studies of official documents, etc. The study of these data allowed us to construct a differentiating table from seven criteria (see Table 5.1).

	Cyclical innovation	Radical innovation
Aim of innovation	First and foremost economic/business focus (standard lexicon: value creation, opening of new markets, new products, being profitable).	Radical innovation focus, strong social dimension (standard lexicon: revolution, society, transformation).
Purposes of the innovators	Willingness for improvement, responding to requirements, solving problems (improving, operating, optimizing, managing, solving).	Willingness to create novelty, having a strong impact (rupture, push, upset, destabilize).
Position facing novelty	Management of uncertainty. Capacity to adapt in a complex and moving world.	Desire for the unknown and search for novelty. Research for potentialities of the unknown, unpredictability.
Position facing an innovation activity	Strong implication. Investment in a project. Risk taking necessary but measured.	Passion, significant risk taking, can continue until being put in danger.
Modes of action	Project mode. Planning, organization, management (managing techniques, method, tools, process).	Trajectory mode. Progress by intuition.
Which guide	Confidence. Flair.	Faith. Vision.
Innovation training	Training in terms of skills to acquire: creativity, management, entrepreneurship, personal development (learning to communicate, to manage a team, etc.).	Training by action. Undertaking personal projects (at school and outside school). Personal development. Immersion.

Table 5.1. *Differentiation of the two families of innovation by seven criteria formulated, starting from the interviews of people in an innovative environment and in engineering schools*

These two conceptions of innovation correspond to two types of innovators. The first, cyclical innovators, position themselves first with the intention of developing an economic market system and producing goods or services in order to create value. The second, radical innovators, are unpredictable in what they produce and in their results, the trajectories for their evolution are unexpected and improbable. They exploit the unknown to develop novelty. Their authors wish more clearly to achieve social change.

Cyclical innovators appear as engineering managers, mastering knowledge, techniques and processes, possessing multiple and often transversal skills capable of adapting to varied situations, and processing vast and complex information. They range from incremental innovations to the marketing of new products coming directly from the application of scientific results. They are situated in a context of uncertainty and probability management. They adopt by experience in their environment, a common professional identity based on the ability to control and deliver their job, which brings them together, characterizes them and legitimizes them. They create value as entrepreneurs or intrapreneurs.

Radical innovators are more extreme in their behaviors. They present very varied profiles, even if we find similar elements in their family context, for example the influences received during childhood. This suggests that similar influences and situations lead to varied and unexpected responses by these future innovators. Their life trajectories are singular, as well as their motivations. Their capacity for innovation is linked to their personality, to the events that they have experienced and to which they have reacted in ways that are specific to them. Their innovator quality comes directly from the evolution of their identity, with which they are confused [LIU 17]. There is construction of a socioprofessional singular identity throughout their life. They are closer to creators/artists, in the sense that their innovations are their "works", more than the projects realized.

Thus, these results for innovation lead us to ask the following questions concerning innovation training: what training is needed for acceptance of innovation?

How can training favor the construction of a singular and unique identity for the innovator (notably for radical innovation)?

How can we integrate innovation training in an established engineering school?

5.3. The two paths for innovation training in professional education

The distinction between the different educational systems does not lie, in our opinion, in didactic and pedagogical methods, but in their foundations: the educational paradigms, i.e. in the conception of knowledge and the teaching theory that are inherent [LIU 15].

Miscellaneous educational paradigms have emerged and have been identified throughout European history, since the teachings of Plato and Aristotle (6th Century BC), up to those of Paolo Freire [FRE 14] and of ATD Quart Monde (fourth world) [DEF 12]. The theoretical study has led us, for the purpose of clarification, to identify two distinct paradigms that shed light on the relationship between types of innovators and curriculum logic. To argue our positions, we took the typical ideals of an educative paradigm [WEB 04]: that of teaching by transmission/acquisition and that of teaching by transformation/emancipation (see Table 5.2).

Criteria	Paradigm transmission/acquisition	Paradigm of transformation/emancipation
Contents	Broken down into disciplines.	General, systemic approach, global approach.
Knowledge	Stable, established, objective, depersonalized and decontextualized concepts	Integrated: multidisciplinary, situated knowledge, evolutive.
Know how	Procedures, tools, instruments, practical skills and techniques.	Open and global policy and/or linked to a diagnosis (clinical policy).
Attitude	Teachable in the form of specific techniques to be acquired.	Attitudes to develop, construction of personality and socioprofessional identity.
Procedure	Chronologically planned, logically articulated.	According to the requirements of the learning activity and the evolution of the learners.
Evaluation	Evaluation chronologically planned, logically articulated.	Formative, autoevaluation, evaluation by peers, based on the learning process.
Teacher/pupil interaction	Authority of the teacher, didactic contract.	Egalitarian and reciprocal, evolving roles of the teaching depending on the nature of the learning.

Table 5.2. *Didactic and pedagogical characteristics differentiating the two educational paradigms*

How can we distinguish these two educational paradigms when we study a teaching method? These curriculums are different because they are inspired by a didactic approach (epistemology of knowledge) and a distinct teaching theory (pedagogical approach).

For transmission teaching, knowledge is divided into different disciplines. The organization for knowledge transfer is planned for using a program. Particular importance is given to the definition of the contents that are composed of knowledge (established concepts, objectified, often decontextualized), know-how (technical and practical skills: what the student must be capable of performing after learning) and social skills (targeting social skills: ability to communicate, animate, lead, manage, etc.).

The relationships between teachers and students are established by a didactic contract. The teacher exercises authority over the student: they are responsible for the lesson. The pedagogical methods are more or less participatory (lectures, projects, group work, etc.). They are all subject to a summative evaluation that must meet certification objectives and meet objectivity and reliability criteria.

Transformation teaching notably involves the construction of a socioprofessional identity. It is based on the transmission of a culture via immersion in an environment and by practice, putting the students into action with time dedicated to reflexivity. Knowledge is situated and evolves. Know-how is declared. The methods include both global approaches, which are redefined along the way when they are no longer pertinent, and a range of precise tools to use according to the requirements. Social knowledge (attitudes, emotions, motivations) is required for training people. It is considered in terms of a person's abilities and skills. Learning depends on the student's commitment, it is their responsibility (self-study). Evaluation is formative, carried out first by the learner in the form of feedback in order to discover difficulties and the steps involved in progression, or by peers sharing their thoughts on the learning policy of each of them.

Relationships between teachers and those taught are reciprocal. They take different forms according to learner identities, those of the teachers, the situation and the context. Education, by its very nature, involves shared learning between these two categories of people, for example teachers also construct their knowledge starting from these very particular situations.

The three fundamental methods for transformation teaching are as follows:

– dialogue: via exchanges and within secured dialogue spaces, in an open and respectful environment, in which students will be able to develop knowledge;

– action: this is at the center of such a method because words can only be situated and come from the reality of the action to have any meaning. Otherwise, the risk would be to fall into theoretical or psychological exchange groups;

– the collective dimension: the production of knowledge, the transformation of the relationship with oneself and with others, in other words the construction of the socioprofessional identity, via the training of a collective where the acknowledgment of alterity leads the engineer to confront and provide questions about themselves and on the relationships that we wish to create with others.

Table 5.2 lists these distinctions as criteria to distinguish these two educational paradigms.

The distinction may seem trickier nowadays, because recently, teaching in general and more particularly higher vocational education have given themselves objectives set out in terms of skills or competences based on the development of missions or projects performed by the students.

But does it make it necessary to adopt the educative paradigm by transformation/emancipation in order to develop identity?

These skills and competences are intended to facilitate acquisition of behavioral skills necessary to fulfill certain defined functions, influencing indeed the learner's professional identity, nevertheless very rarely with an objective to construct an overall professional identity. Furthermore, these projects are often not accompanied by critical thinking or feedback from experience due to a lack of time, willingness or framework in which to do it. However, these processes are essential for innovation training.

What are the innovation training methods in engineering schools? Can we recognize these different paradigms? We have studied in great detail the teaching brochures of around 10 innovation training sessions in order to respond to this question.

5.4. Applied study of the training programs to the innovation of engineering schools

Twelve innovation training courses offered in engineering schools or in the universities[2] (engineering degrees or masters, Bac + 5 level) were analyzed as part of this study. This analysis has been undertaken using the descriptive brochures for each of these training courses on their websites (university year 2015–2016).

Our first observation confirms that there is a complete range of training courses according to the definition that we give to innovation, from incremental innovation to radical or breakthrough innovation, and the others aimed at managing the complexity of the functions required by innovation or what it implies in a company. Indeed, our analysis of the training offered in the engineering schools reveals four types of training with supposed innovation properties:

– innovation management;

– innovation research;

– introduction of new product or services;

– professional training oriented toward the emergence of an innovator identity.

5.4.1. *Training in innovation management*

In this case, the engineer helps to introduce, implement and develop innovation in the company they work for. The engineer has above all a support function for innovation in terms of its technical aspects, but also in terms of management; nevertheless they are not themselves an innovator. These courses approach innovation from the viewpoint of its implementation in the company and from that of its management. From this point of view, their objectives are coherent, and the skills referred to are fully justified. It is regrettable that these aspects are not very explicit in the objectives and that the differences between the professional cultures of exploration and

2 École nationale supérieure des mines, Université Paris-Dauphine, ENS Cachan, ENSAM Paris, École polytechnique, École des ponts et chaussées, Université Paris-Sud, Centrale-Supélec, Télécom Paristech, Université de Lorraine, Université technologique de Compiègne (UTC).

creativity and those of management are not the subject of lessons explicitly included in the presentation of these courses.

We provide the quote to illustrate the objectives of one of these master's courses:

> "to train high level management about technical, financial and human responsibilities, who wish to specialize in the management of innovative projects within existing companies and emerging companies, to create value, to undertake and to be players in the development of a company
>
> To offer this to companies of innovation experts so that the spirit of change can distribute harmoniously in the different departments of the company".

The skills acquired involve mastering tools and management methods for innovative projects. The master's degree offers six modules (360 h) of teaching and a project module to apply the knowledge acquired. Validation is carried out by a written examination based on theoretical knowledge and the presentation of a professional thesis in front of a jury. This master's proposes training by acquisition, which is clearly described and put into practice. It does not explicitly address the development of a professional identity.

5.4.2. *Research training on innovation*

In these training courses, innovation is considered as a new and badly recognized phenomenon, which requires research and studies on both its nature and its practical/professional applications. These courses aim to study the phenomenon of "innovation" based on a scientific methodology. The broad spectrum of the term and the realities that it covers as well as the multitude of dimensions that it touches upon are the difficulties encountered in this kind of training. Also, the three training courses in this category that we have studied approach innovation research in different ways. The first proposes an "approach to innovation processes in their cognitive, technological, organizational, political, social and commercial dimensions". The second aims to "contribute to theoretical and methodological thinking about innovative design and engineering design". More concretely, it involves "performing methodological research allowing us to optimize the design process with application to industrial projects". The third aims to "train the students mastering the operating principles of network industries and the digital economy, equipping them with an analytical and methodology capacity, allowing them in turn to follow strongly evolving phenomena, in the research and higher education sphere".

These training courses consider innovation as an object of study. The content of these training courses emphasizes the research methodology that must confront complexity and disciplinary transversality. In this perspective, the results concerning the resolution of the difficulties encountered by research methods dealing with complex objects appear as a priority and are important. These training courses are the ones that move furthest away from professional innovation training: they appear to offer training for researching the fields of complexity and multidisciplinarity.

5.4.3. *Training regarding the manufacture of new products or services*

In this section, we will cover innovation training as well as innovators. It concerns the training of production engineers who work in new domains, such as computer engineering, or those who improve existing products or manufacturing processes or simply improve them.

The proclaimed ambition of one of these training courses is "to construct a context favorable to the emergence of breakthrough innovation projects, to act as a catalyst for entrepreneurial motivations and prepare students for specific challenges linked to the development of startups". We address two criticisms by studying the curriculum proposed. The first is linked to the underlying design of innovation. It appears as being controllable and foreseeable: it results from the exploration of the potential contained in scientific results; we thus move from science to innovation. This conception reveals that entrepreneurship like innovation is programmable and results from reproducible processes. Risk and the unknown have no place here. Innovation is a project with hazards that are required to be managed, but it is not an adventure.

The second is that the construction of a professional identity is never evoked as a pedagogical objective. Innovation is supposed to appear provided that we place a student in an enabling environment, transmitting to them an amount of knowledge and immersing them in experiences at a pace that does not allow any time for personal assimilation. This approach is related to a conception of training by acquisition rather than by development/transformation, and risks being limited.

These three training courses share an understanding of innovation that identifies them as creating "new" products or services in the sense that they are foreseeable and feasible, but not yet realized, which we call cyclical innovation. While one of them wants to construct a context favorable to the emergence of breakthrough innovation projects, it proposes to its student's dynamics that are described and identified and that risks limiting their imagination and preventing rupture. We could object to these training courses, in that they want to train innovators by transmitting to them known and validated knowledge instead of favoring their singularities, their creativity and their adventurous spirit. They promote cyclical innovation rather than radical innovation.

5.4.4. *Professional training oriented toward the emergence of the innovator identity*

In the case of these training courses, the profession targeted is directly that of the innovator. It assumes that training promotes the construction of a specific professional identity, capable of innovating in every sense of the term, which implies radical or breakthrough innovation. Three programs, which we studied, show this ambition more specifically.

The first is an engineering degree awarded after 5 years that focuses on pedagogy by action and on a solid scientific background. School years with a human dimension, support and self-study combine to create tailor-made paths. The school has a technological platform (FabLab) that allows them to cover all the steps of innovation, from concept to prototype. It has an original department, "behavioral skills/personal development", devoted to the development of managerial skills and personal development, as described: "beyond the know-how and managerial techniques, the department aims to reinforce the students' entrepreneurial spirit, broaden their general culture and their confrontation with human, social and ethical problems, reinforcing their autonomy". The training requires that they take responsibility for the acquisition and application of knowledge to "prepare them to enter into a dynamic of personal development and training throughout their lives". The training also wants "to develop a mindset capable of creating the conditions of autonomy and collective responsibility".

This teaching is based on numerous internships with different aims: cultural (first year: linguistic training), social (second year: work training) and professional (third and fourth years: industrial project; fifth year: final assignment).

This training is open to allowing the choice of a singular route, as detailed in the school's presentation. The testimony of a former student details the purpose for which the course is given: "some very original courses will allow us to discover what our profile is from a psychological point of view". It is oriented toward student awareness of "who they are and what they want. It is a good-quality school and is well suited to those who are ready to learn more about themselves"[3].

The path of another of these programs is positioned at the intersection of the creative, cultural and new industries utilizing digital techniques that establish and increase the imagination (cinema, video games, software industries, 3D, ITC, etc.) in the new industrial environment of intensive innovation that is emerging. It values both the analytical and creative dimensions of the innovation actors. The challenge of this training is to articulate the three parts (engineering, human sciences and design) in a transdisciplinary way to prepare the students for industrial jobs requiring creativity and a capacity to seize the innovation dynamic. Its teaching shows an orientation toward a global and transdisciplinary understanding of knowledge, a methodology based on the open and evolving formalities to measure whether their progression causes new problems. Furthermore, provisions such as appropriation, thoughts on dynamics, understanding of the cultures of the human sciences, the effort toward transdisciplinarity and cooperation are present and involve the development of the person and identity among learners.

These training courses respond to criteria that we have defined as being from an educational paradigm of construction of a socioprofessional identity.

---

3 http://www.ciao.fr/ENSGSI_Nancy__Avis_877692.

5.5. What innovation training should be integrated in an engineering school?

This empirical study corroborates the idea that institutional innovation training has multiple dimensions and is different according to the meaning that we give to the term innovation.

It also strengthens the response to our question about training in radical innovation that will be based on the construction of a singular and specific identity for each innovator.

This study has however raised new questions. We can see that the three last programs studied are marginal vis-a-vis their institution: engineering degree and master's degrees are offered in universities, the third master's degree belonging to a major school has been stopped after its second year in existence. This raises a question about the transformation required by the introduction of an educative paradigm for identity construction within the schools based on an educational transmission paradigm. Within an institution, can an individualist and hierarchical culture coexist with an egalitarian and emancipating environment?

Some schools who are aware of the challenge are trying new pedagogical methods.

This is the case for 42, a school without professors or courses, open 24 h a day and breaking with the "traditional" schools. Its teaching method involves the immersion of the students in a distinct environment and undertaking projects having progressive complexity. The creation of 42 is closely linked to the evolution of Epita, a private engineering school specializing in "computing" (computing and advanced techniques school) belonging to the IONIS group. In 1999, Epitech was created by the same group. Epitas' little sister is distinguished by its pedagogy, which is highly focused on "projects" and the fact that it directly provides training in programming, excluding nearly all general lessons. According to the founders, Epitech was created when Epita changed its pedagogical method in order to ultimately obtain acknowledgment and validation of the degree by CTI. Epita had implemented rather atypical pedagogical methods at the time: active pedagogy, tutoring between students, etc. Some elements, such as the correction of homework by more experienced students, have had to be abandoned in order to fit the commission's criteria. Its first director says he expected to develop Epitech toward more active and less transmissive teaching. Indeed, there are very few courses (mostly language courses). Nevertheless, a rivalry between the management teams has quickly established between Epita and Epitech: personal conflicts, but also different teaching visions. Epitech, which is accountable to the group, is constrained in its willingness to go to the very end of a need for pedagogical transformation. At the end of 2012, part of the Epitech team decided to establish its own structure; 42 was created, financed by a patron and self-proclaimed "code breaker".

Thus, it is by the successive creation of new structures that the pedagogical model of 42 has in the end emerged, freed from institutional constraints in terms of teaching.

Each in their own way, the three types of training courses that we have classified in the category "oriented toward the emergence of an innovator identity" and the 42 school meets the criteria that we have defined as being those of an educative paradigm involved in the construction of a socioprofessional identity. They are, similar to the creation of an emancipating environment[4] for their learners, characterized by:

– a large choice of trades or professions in which to engage and continue development following the training;

– the capacity to have a global, complex and concrete vision of situations;

– self-study by action and cooperation linked to critical thought;

– the definition of formalities during which the journey is progressively constructed;

– the development of a singular personality and a confident and voluntary identity, capable of listening to intuition and engaging in an assumed unconventional behavior.

The notion of an emancipating environment highlights several key elements that we will now detail.

4 We have defined and experimented with since the end 2016 such an environment within a training program [VER 17].

5.5.1. *Emancipation promotes engagement in learning*

The commitment to learning is the question at the core of all methods aimed at education: a black box, to which all teachers are confronted, whether initial or continued training, other or self-training; the most efficient didactic method in the world will always meet the students' willingness to learn or not to learn. The notion of emancipation proposes a path of explicitness residing not only in the perspective of the students alone but placing the position of the educators themselves at the center of the method to allow for commitment. They renounce their ability to provide knowledge and to organize training. They accept the subject learner's freedom, in taking the risk of being rejected, but it is the acceptance of the risk, which is precisely the condition for freedom of commitment.

5.5.2. *Importance of creating an environment rather than a semiclosed method (type of school)*

The application takes place in the real world, and this action is transformed into knowledge during dialogue sessions, which represent the core training, where speech is the fundamental tool. As P. Freire says: "we discover in speech two dimensions: action and reflection, solidary in an interaction so absolute, that by deleting one of them, the other suffers immediately" [FRE 14, p. 71]. The training space, by welcoming words on action, reinterprets it, explains it and analyzes it in a learning perspective and improve its effectiveness.

5.5.3. *The burden of the transformation identity by action is supported by a collective*

All training methods revolve around real projects that are chosen and undertaken by students. They are not inserted in a learning context with artificial consequences. The role of the created collective is fundamental to accompanying immersion in the real world. This requires a specific teaching position: its first role is to accompany the meeting (in the sense of Freire) of the individualities, helping to weave the links of the collective to come.

This radically differentiates the methods of this paradigm from the tutoring logic where the students are accompanied by a teacher who is not always trained in this type of relationship. D. Lemaître shows that "the tutoring method leads inevitably to a psycho-sociological transmission" [LEM 10], leading to confusion of training logics. He emphasizes the importance of explaining the training objectives when going from a didactic logic to a project of accompanying identity. With regard to an emancipating environment, the guarantee of the learning environment is the production of projects with a collective of dialogue always focused on action. This allows us to avoid falling into improvized therapy sessions.

5.6. Conclusion

The study undertaken has allowed us to provide the following results:

– innovation is a term that covers processes belonging to different dynamics. With respect to innovation training, we have proposed organizing these dynamics into two families: cyclical innovations and radical innovations;

– the training for each of these families is based on different educational paradigms: the educational paradigm by transmission/acquisition for cyclical innovations and that of transformation/emancipation with respect to radical innovation.

The first experiences indicate that training in cyclical innovation can be integrated into the engineering school curricula as they currently exist with their traditions, but that training in radical innovation often collides head on with these traditions and requires structural change within organizations, as well as their cultures, notably:

– the transformation of the semiclosed method of an established organization to that of an emancipating environment as we have defined above. The training must leave the educative institution to take place within the activities of the world with economic, social, technical and real issues. We return to the educational institution only for the evaluation of the experience acquired during these experiences to prepare the knowledge;

– the transition from a hierarchical culture with predefined roles to a culture of dialogue between diverse actors involved in teaching. Teaching is no longer the transmission of knowledge, but a collaboration of knowledge by each participant, both for students and professors;

– the change in roles and attitudes of teachers, who as providers of knowledge must enter into a process of shared learning with their students (each participating with different levels and objectives);

– knowledge is no longer archived and theoretically formatted, but dialogical knowledge that adapts and recreates in situations to understand, formulate the problems that they have, and invent paths for their resolutions.

It is thus, in other words, transforming a culture centered on the transmission of knowledge by a culture centered on action and dialogue between learners and teachers, with a view to allow learning to emerge from experience and collective intelligence.

5.7. References

[ALT 02] ALTER N., *L'innovation ordinaire*, PUF, Paris, 2002.

[BER 13] BERGSON H., *L'évolution créatrice*, PUF, Paris, 2013.

[CHR 13] CHRISTENSEN C., DYER J., GREGERSEN H., *Le gène de l'innovateur*, Pearson France, Montreuil, 2013.

[DEF 12] DEFRAIGNE-TARDIEU G., "Freire, Alinsky, Dolci, praticiens de l'émancipation", *L'Université populaire Quart Monde : La construction du savoir émancipatoire*, Presses universitaires de Paris Ouest, Nanterre, 2012, available at: http://books.openedition.org/pupo/2390.

[DES 12] DESCHAMPS J.-P., "L'innovation et le leadership", *Revue économique et sociale*, no. 4, December 2012, available at: https://www.bcge.ch/pdf/2012_moteurs_innovation_deschamps.pdf.

[DIC 62] DICTIONNAIRE DE L'ACADÉMIE FRANÇAISE, 4th edition, 1762, available at: http://atilf.atilf.fr/dendien/scripts/generic/cherche.exe?15;s=4161210615.

[FRE 14] FREIRE P., *Pédagogie des opprimés suivi de Conscientisation et Révolution*, François Maspero, Paris, 1974.

[GAG 11] GAGLIO G., *Sociologie de l'innovation*, PUF, Paris, 2011.

[LEM 10] LEMAÎTRE D., "Construction des identités et projet de formation", in RAUCENT B., VERZAT C., VILLENEUVE L. (eds), *Accompagner des étudiants*, De Boeck Supérieur, Louvain-la-Neuve, pp. 87–107, 2010.

[LIU 15] LIU T., "Quelle pédagogie pour former des innovateurs?," *Actes du colloque QPES Innovation dans l'enseignement supérieur*, Brest, France, 2015.

[LIU 17] LIU T., "Qu'est-ce qu'un startuper ?", *Agora Débats/Jeunesses*, vol. 751, no. 1, pp. 117–130, 2017.

[MEN 11] MENISSIER T., "Philosophie et innovation, ou philosophie de l'innovation?", *Klesis – Revue philosophique*, vol. 18, pp. 10–27, 2011.

[SCH 04] SCHUMPETER J.A., *Histoire de l'analyse économique*, Gallimard, Paris, 2004.

[VER 17] VERZAT C., PETIT F.-X., LIU T., "Notre modèle est ouvert et centré sur l'humain", *Entreprendre & Innover,* vol. 31, pp. 36–42, 2017.

[WEB 04] WEBER M., *L'éthique protestante et l'esprit du capitalisme*, Gallimard, Paris, 2004.

6

Innovation Training and Entrepreneurship in French Engineering Higher Education Institutions: An Investigation of the Commission des Titres d'Ingénieur

6.1. Introduction

Accreditation agencies now play an important role in the design of training programs through the guidelines that they produce and the evaluation procedures that they undertake periodically. In France, this is particularly the case for the *Commission des Titres d'Ingénieur* (CTI) that, since its creation in 1934, has historically contributed to structuring and supporting the development and evolution of engineering education in France via the engineering degree (master level 2). In line with France's ambition to create an engineer with a broad range of skills, beyond only technical specialization, CTI's guidelines expressed in its document *Références et orientations* (Standards and Guidelines) [R&O 16] cover all domains that must control at a minimum an industrial company executive, within the different topics of management and social and environmental awareness.

This is why CTI has since 2012 adopted the theme of innovation, for which the stakes have become significant with regard to engineering education. Within the guidelines addressed to schools, there is a specific chapter on this issue, which is now included in the points addressed during

---

Chapter written by Anne-Marie JOLLY and Julie NOLLAND.

the evaluations. Innovation is explicitly linked to the theme of entrepreneurship, with the aim of encouraging graduate engineers to create new activities and to stimulate industrial renewal. As an observer of practices, CTI has since initiated an investigation in order to better understand how higher education institutions integrate these incentives, and what they actually implement in the field.

The results of this investigation are presented in this chapter as a general overview of the observable pedagogical practices. After a brief presentation of CTI and its specific expectations with regard to innovation training, this chapter presents the approach adopted for the investigation and an overview of the responses provided by the higher education institutions.

6.2. Study context: CTI and engineering higher education institutions

In France, to obtain the title of graduate engineer outside the framework of *Validation des Acquis de l'expérience*[1], it is required to follow a curriculum in one of the 206 engineering higher education institutions issuing a CTI-accredited degree program. At the end of 5 years of post baccaluréat studies, the engineering students obtain the title of graduate engineer from the higher education institution in which the training has been undertaken, leaving with a master's degree.

The specific feature of CTI, a quality assurance agency and accreditation body, is its parity composition with half representatives of the academic world and half from the socioeconomic world: 16 members are from higher education institutions, eight members from professional organizations and eight members from trade unions.

This parity provides the commission with unique guidance with regard to quality assurance, quoting, for example, the predominant place given to employability in its evaluation approach and great concern about the new needs arising from societal changes, such as environmental awareness or

1 http://www.vae.gouv.fr. This *validation* is a process of taking into account formal and/or informal learning in order to award the degree. It is based on a review of professional activities and experienced gained during practical years. The procedure was implemented by the law of social modernization, dated January 17, 2002; its implementation is mandatory in all French institutions.

ethical dimension. All CTI members, gathered in plenary session, state the accreditation of a program at the end of an audit process, including the drafting of a self-evaluation report by the higher education institution and an on-site visit, every 5 years. Its members also approve every development of the standards and guidelines, including evolutions of accreditation criteria, and sometimes even new criteria.

The accreditation criteria cover various fields, available in a public document called *Références et orientations* (Standards and guidelines) [R&O 16]. These criteria are revised regularly (every 3–4 years) to respond to societal changes. This thought process and continuous improvement, based on the reference documents, is also a requirement of the ENQA (European Association for Quality Assurance in Higher Education), "the agency of agencies". This European body periodically undertakes audits in the quality assurance agencies of the European higher education area (EHEA) in order to check the conformity of their procedures to European Standards and Guidelines [ESG 12].

CTI has been a member of ENQA since 2005 and has been registered since 2010 in the EQAR (European Quality Assurance Register for Higher Education), identifying all the quality assurance agencies of the EHEA meeting these common criteria.

Beyond changes in its criteria, including those regarding innovation and entrepreneurship in 2016, CTI considers that certain themes require increased awareness by the institutions. CTI thus expects a clear evolution of the curricula and the current practices in the institutions with accredited programmes. This is confirmed first by the specific data requested from them, but the commission can also ask, in addition to the self-evaluation report, other documents such as the "focus", a procedure upon which the analysis developed in this chapter is based.

6.3. Expectations regarding entrepreneurship and innovation

CTI is concerned with current societal problems and particularly those raised in the European program H2020 [EUR 06, EUR 16] has reinforced its expectations regarding openness to innovation and entrepreneurship.

CTI associates the problem of innovation with that of entrepreneurship, considering that this link is relevant in a didactic perspective within educational activities.

Evaluation of the openness to innovation must be supported by concrete productions, verifiable within institutions, which will allow the development of a state of mind and the implementation of specific activities and projects. This openness also results from a suitable pedagogical approach.

According to the CTI, this learning process must encourage critical thinking, the right to make mistakes, risk taking and support the expression of creativity while allowing the engineering student to become aware of user needs: customer orientation must be at the core of the creation process.

The pedagogical capitalization of experiences, transdisciplinarity, curiosity and openness to the world are consequences of this process.

This CTI initiative seems to bear fruit. For example, since 2016 "certified data", indicators filled in yearly by engineering higher education institutions and which are part of the self-evaluation report show that 182 institutions have declared training students in these fields.

Recall that the indicators requested cover 11 main themes

(12 from 2018 onwards) listed in the following:

– general information about the institution;

– general information about the programs (workload, ECTS, etc.);

– number of graduates (by program);

– research environment;

– recruitment;

– social openness;

– innovation; valorization;

– relations with companies;

– internationalization of engineering programs;

– employment;

– student life; reputation;

– quality assurance (starting from 2018).

The certified data are a key element for the institution's self-evaluation report and a fundamental tool for the audit panel. They are certified as being accurate by the dean of the institution and published on the CTI and institution websites, thus providing more information for the general public.

The section of indicators dedicated to expectations regarding entrepreneurship and innovation, "VI. Innovation-valorization", includes the following items:

– VI.1.a: there is specific teaching for all the students about entrepreneurship and innovation management in the institution (response expected: yes/no);

– VI.1.b: there is specific teaching for all the students on entrepreneurship and innovation management in partnership with the institution (response expected: yes/no);

– VI.2.a: there is an incubator in the institution (response expected: yes/no);

– VI.2.b: there is an incubator in partnership with the institution (response expected: yes/no);

– VI.3: number of engineers coming from the institution supported in business creation by an incubator in the past 5 years (response expected: quantified indicator);

– VI.4: the institution is connected to a Pôle étudiant pour l'innovation, le transfert et l'entrepreneuriat (PEPITE)[2] (response expected: yes/no);

– VI.5: number of students holding the status of "student entrepreneur" (response expected: quantified indicator).

CTI reasons in terms of expected skills. Among the 14 key skills expected from French engineers, we find "the ability to undertake and innovate in the framework of personal projects or by using initiative and involvement within the company in entrepreneurial projects" (R&O, section IV.2, competence 12 [R&O 16]). It is certainly a "shy" mention, but one

2 Student center for innovation, transfer and entrepreneurship - https://www.pepite-france.fr.

which does not currently appear in most other reference frameworks, notably that of the European Network for Accreditation of Engineering Education (ENAEE) [ENA 15], which promotes the EUR-ACE® label.

Among the skills targeted by engineering training, the most directly affected by this innovation training process are from CTI's point of view:

– scientific and technical skills: designing and leading an innovative company project;

– relational and managerial skills: anticipating and making decisions in uncertain situations;

– cognitive skills: solving problems in a creative, inventive and innovate way;

– transverse knowledge: entrepreneurship, intellectual property.

This training process is generally implemented at two levels: general training performed at the beginning of the curriculum in order to "let the seed sprout" for all students and a possibility of providing a more in-depth training performed with partners of the institution.

The institution is free to set up transverse activities that it finds most appropriate. We can quote, for example, the production of a personal or collective project of entrepreneurship/innovation seeking the skillset out above: listening to customer needs, displaying creativity, experimentation, production of a business plan, etc. This kind of project requires time dedicated to cooperation with other specialties or fields as well as personal work.

Feedback on the devices implemented and the performance indicators chosen must be planned by the institutions, according to CTI expectations.

CTI is also interested in the engineering higher education institutions' participation in PÉPITE [PÉP 14, PÉP 16] implemented by the French state in September 2014, allowing access to the status of "student entrepreneur" and the "national diploma of student entrepreneur D2E". The French government, wishing to develop entrepreneurship in universities and schools, has provided students with the opportunity to develop their company project (staging, coaching) while undertaking their schooling. In September 2016 [DC 16], 156 engineering higher education institutions were deploying their actions on entrepreneurship in connection with one of the 29 regional PEPITEs and 360 engineering students held the status of student entrepreneur.

Among the documents studied during the audit, including both the file supplied by the institutions and its appendices (studied prior to the visit and during the consultation of documents during the visit), particular importance is given to information showing that teacher–researchers are associated with the strategy of openness to innovation and entrepreneurship, as well as the integration of the institution in a collaborative and partnership ecosystem with the socioeconomic world.

6.4. Investigation conducted within engineering higher education institutions (called "Focus")

In February 2016, an investigation was launched to develop various topics mentioned or developing in the CTI reference framework, namely innovation and entrepreneurship, health and safety at work, sustainable development and societal responsibility.

Only the institutions visited in the framework of a periodic audit had to mandatorily write a focus on these themes, a voluntary contribution being possible for the other institutions.

The need to know in more detail the practices of institutions regarding preparation for innovation and entrepreneurship led the CTI to explain further, in this focus procedure, the elements requested in the accreditation file.

The compilation of answers from around 10 institutions of different sizes and specialized fields provided CTI with an insight into current institutional practices and facilitated identification and compilation of the best practices. The goal is that the institutions short of ideas to implement can seize these practices and adapt them to their own specific reasons.

The questions addressed to the institutions are presented in Box 6.1.

Cti
Commission des titres d'ingénieur

FOCUS

INNOVATION AND ENTREPRENEURSHIP (I&E)

Entrepreneurship

Objective: At a minimum, the institution informs the students about entrepreneurship (while promoting business creation) and can be part of a territorial ecosystem for business creation. Beyond this, the institution provides training about business creation by deploying internal means, or by providing access to external training provided by the ecosystem (for example in connection with D2E).

Information and questions addressed to the institution:

– Business creation awareness:

- Describe the means for raising awareness regarding business creation implemented in specifying the leader, for operations in partnership with other institutions of the ecosystem. Also note that the target audience is the students (option, all the students);

- which good practices have been identified?

- how can we measure the impact of actions on the entrepreneurial spirit of students?

– Training in business creation:

- in the case where training exists in-house: describe it by mentioning the intervention of territorial ecosystem partners involved in business creation;

- what seems to be the correct model for this kind of training in an engineering higher education institution?

- in the case of total or partial outsourcing of training: describe the conventions established with partners, notably with regard to the D2E;

- has a good practice been shown via the implementation and the management of partnerships.

– End of studies internship dedicated to business creation (PEPITE):

- describe the promotion of the internship formula and the questions related to its implementation;

- what role can the institution in the PEPITE territorial system play? What good practice has been noted in the management of this kind of partnership?

Innovation

Objectives:

– valorization of an experimental innovation practice, including an exercise in creativity and a pedagogy of the economic and social evaluation of innovation;

– investigate pedagogical capitalization.

Information and questions addressed to the institution:

– Experiments (project activities) of innovation:

- describe the activity performed and the good practices observed. Describe the specific pedagogy, its collective implementation and its career development prospects;

- how can we capitalize on the pedagogical experience?

- how can we evaluate the impact on students in practicing such experiments?

- in the current offer in management of innovative projects, which theories, concepts, methodologies can represent a basis for educational support?

– Creativity:

- provide a brief description of the activity performed and the good practices observed.

– Pedagogical activities highlighting the economic, social and societal contribution of innovation:

- describe these activities and their link with other pedagogical activities. Detail the questions linked to the valorization of innovation and its ability to answer customer needs;

– FabLab, innovation ecosystem:

- if the institution has a FabLab, describe its strengths and weaknesses. Please do the same if the institution is involved in a partnership established with an external FabLab;

- describe the institution's involvement in its innovation ecosystem. Describe the creation of very positive partnerships, which have a real measurable impact on the pedagogy and the validation of the institution's activities.

Box 6.1. *Focus questionnaire presented to engineering higher education institutions*

It should be noted that in the initial implementation of PEPITEs, these elements that the focus wish to present were part of the activities planned with regard to the PEPITEs themselves.

This was the first time that the CTI has undertaken such an investigation with institutions and the answers provided proved to be educational and very encouraging for the development of innovation training and entrepreneurship.

6.5. Answers from the institutions

CTI has therefore collected 12 descriptive documents ranging from 4 to 10 pages helpful for understanding the involvement of institutions in innovation and entrepreneurship.

On the topic of entrepreneurship, several institutions have contributed voluntarily, and it appears clear that the subject is taken into account in institutions, most of the time from a date prior to the introduction of these concerns in CTI's references and guidelines.

However, according to the legal status of institutions, the definition of innovation and its inclusion are rather variable. The actual link between the institution and research depends indeed a lot on this status. Regarding entrepreneurship, the implementation of PEPITEs has certainly contributed to the generalization, even to the normalization of this concept in institutions.

In order to classify the answers, it appeared interesting to CTI to distinguish several complementary fields. Some fields can certainly be interconnected, but this classification has the advantage of making pedagogical innovations more visible. Here is the list of the fields that we have identified to structure our analysis grid:

a) the pedagogical organization in the sense of:

– the decision to teach a particular course in order to acquire a particular competency,

– the level of study addressed by the teaching,

– how the training topic interferes with the specific skills area of the institution,

– the way to teach these courses toward a concrete or more abstract learning,

– the staff participating in the acquisition of this learning: institutions or external staff;

b) the material and human resources made available: creation of dedicated zones within or outside the institution (FabLabs, co-working spaces, etc.);

c) partnerships.

This organization is generally only possible thanks to the presence and availability of technical centers or platforms facilitating its implementation. Entrepreneurship being followed closely by the territorial authorities, a clear expression of their support allows, with regard the analysis, the possible definition of a correlation between this support and the results obtained at the end of the process.

This classification is used in the rest of the chapter; in section 6.7, we further explain about the resources and external partnerships. Furthermore, a broader vision of the environment of this investigation can be consulted in [JOL 18].

6.6. Pedagogical organization

Many institutions consider that the appropriation by the students of innovation and entrepreneurship has to be led gradually [VER 03]: indeed, not all of them are intended to immediately create a company. Two phenomena speak in favor of such an organization:

– the training of engineers "*à la française*", including at least one-fourth of human and social sciences, provides a teaching style leading to skills or to a state of mind that can foster innovation and entrepreneurship. We can mention among other things: self-awareness, autonomy, taste in risk-taking and risk management, self-confidence, creativity, good interpersonal skills, the capacity to work in a team and to manage, networking, project management and associated resources, knowledge of management tools, etc. This implies that some of these skills are acquired *de facto* by any engineering graduate who may not have to use them immediately;

– the long-term effect: some engineers, during their professional life and at various times in their careers, could use these skills for their employer, or decide one day to establish their own company. It is therefore necessary to sow this "seed" during the training.

According to the institutions, these teaching actions occur at two levels: on the one hand, to identify the potentially involved students and, on the other hand, to allow them afterwards to develop their project. A rather widespread organization consists of separating the *information* level (*arousing the appetite*), using round tables or times for exchange, providing no new competence, from the *innovation awareness* level (developing entrepreneurial skills). This module can be mandatory, lasting from 12 to 24 h, which allows the production of creative games, creativity workshops and any activity completing the skills already developed in human and social sciences teaching. Nevertheless, in these two phases, the possibility offered to students to exchange is a key element. This organization has an impact on the time granted to entrepreneurship teaching referents by their management. Encounters, such as "live my life", allow a student to share the daily life of a contractor (most of the time a graduate from the same institution) that can be positioned at two levels.

Access to the *student entrepreneur* diploma and to the PEPITE device is beyond these levels, it is really about *taking action*. Resources are developed individually toward the specific project of a student with coaching adapted to this project. It is also within this context that links with the ecosystem can be deployed.

From a pedagogical point of view, active pedagogies are unanimously recommended, unlike lectures about entrepreneurship. Whether through serious games, competitions (team trophies at the scale of a territory), among others, the playful approach is rather shared. These "learning by doing" practices also provide the possibility of immediate feedback, which is very useful when playing down mistakes. Training and implementation of *design thinking* contributes, along with transdisciplinary teachings to openness for innovation. Finally, bibliographic "miniprojects" can be considered as useful to understand innovation, especially jointly with information regarding intellectual property.

A rather widespread practice present in institutions is the challenge "48 h to innovate". Initially developed by a specific institution [INP 11] whose core business and research topic is innovation, this project has spread in France and internationally; it involves students and teachers coming from various academic backgrounds. The timing and specifications are strictly defined.

Nevertheless, these projects remain the most common pedagogical practice to develop innovation. They can take shape of a short-term task in a company. Internships in start-ups are also offered by some institutions, as well as the oldest student form of entrepreneurship, the *Junior Entreprise*[3], which is currently experiencing a revival.

However, these practices must always fit into a new mindset by the teachers. Indeed, it would be contradictory to seek to develop student creativity and their right to make mistakes without mobilizing these elements in the pedagogy of other teaching (besides entrepreneurship): this could even be confusing for students and their training experience.

It is rather new in France to foster teaching and research staff to create startups or other companies, especially in the public sector. Still, in other areas than engineering (law for example) these practices have long been integrated. CTI has also observed such practices among researchers during assessment visits abroad (especially in Flanders or in Switzerland). A cultural (r)evolution is undoubtedly in progress in France at this level.

3 https://junior-entreprises.com.

6.7. Resources implemented and external partnerships

As stated previously, the time available for teachers and referents is a key element of success. The availability of dedicated premises allowing co-working to take place within the institution can be a key element, for example for company presidents. The co-construction, with an external structure, of a department dedicated to a specific activity theme can be a trigger (typically the platforms located within institutions are used for smaller scale innovation projects on specific topics). Students can also come out of their environment and face the corporate world and better understand its issues. The presence of a FabLab in the institution is an important element because it can be a trigger for some innovations (if students have full access). According to the focus, FabLabs have developed in almost all of the institutions surveyed. This is more widespread in some fields: mechanics, ITC, biomedical, electronics, etc.

The incubator housed inside the institution (there are 146 in French engineering higher education institutions [DC 16]) has the advantage of reminding students that the creators on site could be one of them; their great interest also lies in the proximity with research laboratories and some approach innovation through this angle.

We can clearly distinguish in the focus the communities for which innovation and entrepreneurship are seen as key elements to boost the territory and companies: distribution of awards during entrepreneurship competitions that they have created, scholarships and the achievement of challenges enhancing emulation between institutions is fairly constant.

The institution has generally acquired its platforms thanks to the support of communities willing to strengthen the institution–company link. This is particularly documented in the metropolitan areas or the territories where this obligation to innovate is most clearly expressed, either because of competitive issues with the neighboring region or because of a decline in historical activity and thus making room for new horizons.

The existence of technical research centers (such as the *Centre d'études technique des industries mécaniques*, CETIM) within an institution is also a good breeding ground for innovation, since these centers are well aware of the projects of the companies involved. This organization allows companies to build a real proximity with the students and to implement realistic institution company projects.

The development of innovation through undertaking internships in research laboratories in France seems currently less obvious. However, the realization of such internships abroad (USA or Flanders for example) in laboratories creating startups could be more decisive: for the time being, researchers do not seem to be sufficiently involved in the startup creation matter.

Some institutions work together with a management or design institution for example, and this transdisciplinary approach leads to very beneficial conditions, especially in the framework of common projects between the students.

Nevertheless, indicators about business creation communicated by the institutions, namely 2,157 companies created by engineering student over the past 5 years [DC 16], are not initiatives. The number of companies still operating by the end of the next 5-year cycle is a solid indicator of the shift taken by the institutions.

6.8. Conclusion

The high proportion of human and social sciences in the training of French engineers guarantees their ability to complement their skills toward entrepreneurship, if they wish. Nevertheless, in view of the survey results, a change of mentality in the institutions still seems necessary:

– for the students, in order for them to become aware of the interests of innovation and accept it as a necessary starting point for creativity-related exercises;

– from the teaching staff, in order for its behavior to be really motivating for the students.

Furthermore, with respect to specific pedagogical practices, *Design Thinking* seems to be under implemented in the teaching provided: this technique, widespread in design institutions, allows confidence levels to increase for some students in engineering education institutions and leads to breakthrough innovation [BRO 14].

The indicators for *student entrepreneurs* (currently 360 [CTI 17]) communicated by the institutions are rather high: it is nevertheless important to check, at the end of a 5-year cycle, the number of sustainable companies.

The signs of an actual recognition of innovation as a training objective are less convincing so far. France seems to struggle to fully take this opportunity. This can be explained by a historical sequentiality: before 2008, institutions fostered innovation through their research laboratories, and then societal developments have shifted attention toward the more immediate concern of entrepreneurship. We can bet that, in the third phase, a balance between these two aspects will be met.

6.9. References

[BRO 14] BROWN T., *L'esprit design: Comment le design thinking change l'entreprise et sa stratégie*, Pearson, Montreuil, 2014.

[CTI 17] CTI, "Données certifiées de la CTI", *Colloque de la CTI*, Paris, France, 2017.

[ENA 15] ENAEE, EUR-ACE® Framework Standards and Guidelines, instructions, 2015, available at: http://www.enaee.eu/wp-assets-enaee/uploads/2015/04/EUR-ACE-Framework-Standards-and-Guidelines-Mar-2015.pdf

[ESG 12] ESG, Standards and Guidelines for Quality Assurance in the European higher education area, instructions, EURASHE, Brussels, 2012, available at: http://www.enqa.eu/wp-content/uploads/2015/11/ESG_2015.pdf.

[EUR 06] EUR-LEX, Promoting Entrepreneurship in School and Universities, European Commission memorandum, 2006, available at: http://eur-lex.europa.eu/legal-content/EN/TXT/?uri=LEGISSUM%3An26111.

[EUR 16] EUROPEAN COMMISSION, EACE/EURYDICE, Entrepreneurship Education at School in Europe, Report, 2016.

[INP 11] INP, "48 heures pour innover", *iCré@-Formation*, 2011, available at: http://icrea-formation.inpl-nancy.fr/resultat_48h.html.

[JOL 18] JOLLY A.-M., LEGER C., "French Engineering Universities: How they deal with entrepreneurship and innovation", in AUER M., KIM K.-S. (eds), *Engineering Education for a Smart Society, Advances in Intelligent Systems and Computing*, Springer International Publishing, Cham, pp. 284–294, 2018.

[PÉP 14] PÉPITE, "Statut d'étudiant-entrepreneur", video, 2014, available at: http://www.dailymotion.com/video/x1zwdlk_statut-d-etudiant-entrepreneur_tech.

[PÉP 16] PÉPITE, France : le site web des étudiants entrepreneurs, 2016, available at: http://pepite-france.fr.

[R&O 16] R&O, *Références et orientations*, Commission des titres d'ingénieur, available at: https://www.cti-commission.fr/fonds-documentaire/docu-ment/15, 2016.

[VER 03] VERZAT C., FRUGIER D., BACHELET R., *et al.*, "Helping engineers to become entrepreneurs; attitudes, behaviors, beliefs, skills: what are the educational factors of their entrepreneurial spirit?", *Int Ent 2003 Internationalizing Entrepreneurship Education and Training*, Grenoble, France, September 8–10, 2003.

7

Determinants of Skill Matching among Young Hungarian Engineers

7.1. Introduction

Problems around matching employees' skills and competencies to the job requirements have always been one of the structural characteristics of the labor market in advanced economies. Following the global economic crises, in 2008/2009 skill matching again came into focus. It deserves special attention in the case of highly skilled professionals with specific competencies, since accumulating complex and specified skills requires serious investment from both the individual and society as a whole. Matching skill supply and demand is essential in making these investments profitable and in efficient and sustainable utilization of individual and social resources.

Engineers are in a specific position within the intellectual professionals. They play a crucial role in technological progress, innovations, adaptation of rapid technological changes and in the contribution to the material wealth of societies [CLA 10]. With respect to their skills, there are two critical issues to be tackled. The first one refers to the extent engineers are prepared to fulfill the labor market requirements and are concerned by skill gaps and deficits and the second refers to the extent their skills are utilized. Skill gaps and deficits are barriers to competitiveness and economic growth, while misutilization or underutilization of engineers' skills can result in lower satisfaction with work, a decrease in productivity and motivation loss

Chapter written by Péter CSIZMADIA and Zsuzsanna VEROSZTA.

[CED 10]. Skill mismatch can be traced back to several different sources, from poor or inadequate training to the inefficient skill utilization practices of the companies employing engineers.

This chapter seeks to analyze the most important individual and institutional factors determining skill matching in the case of newly graduated Hungarian engineers. The central issue is what individual and institutional factors account for their perceived mismatch between the acquired and required skills and to what extent structural characteristics of engineering higher education and workplaces account for proper skill matching.

7.2. Theoretical background

Skill matching is a complex term that covers heterogeneous types of imbalances between skills acquired by labor and job requirements [ILO 13]. Skill matching simultaneously refers to the skills individuals possess, the skill provision of the educational system and the utilization of skills by the organizations. There are no statistical measures that can capture the complex character of skill matching, therefore in most cases proxies are used to describe its various dimensions [ILO 13]. For instance, qualification, the length of education (measured in years) or skills are often treated as equivalent measures of skill supply, although this is definitely not the case. If someone has a high qualification (PhD for instance) and works as a university lecturer, his or her qualification matches his or her job, but if he or she lacks proper communication skills, he or she is obviously underskilled in that position where communication is a must. Table 7.1 provides a brief definition of the most frequently used terms concerning skill matching in order to create a better understanding of the various dimensions of the phenomena.

Overeducation	A situation in which an individual has more education than the current job requires (measured in years).
Undereducation	A situation in which an individual has less education than the current job requires (measured in years).
Overqualification	A situation in which an individual has a higher qualification than the current job requires.
Underqualification	A situation in which an individual has a lower qualification than the current job requires.
Overskilling	A situation in which an individual is not able to fully utilize his or her skills and abilities in the current job.

Underskilling	A situation in which an individual lacks the skills and abilities necessary to perform on the current job to acceptable standards.
Skill surplus	A situation in which the supply of available people with a particular skill exceeds the demand for it.
Skill shortage	A situation in which the demand for a particular type of skill exceeds the supply of available people with that skill.
Skill gap	A situation in which the level of skills of the currently employed is less than that required to perform the job adequately or the type of skill does not match the requirements of the job.
Skill obsolence	Skills previously used in a job are no longer required and/or skills have deteriorated over time.
Vertical mismatch	The level of education or qualification is less or more than required.
Horizontal mismatch	The level of education or skills matches job requirements, but the type/field of education or skills is inappropriate for the job.

Table 7.1. *Frequently discussed terms concerning skill matching*[1]

In this contribution, we focus on the issue to what extent young engineers and IT specialists use the knowledge, skills and competencies acquired during their studies in their current workplace, which is related to the term "horizontal matching".

With regard to the employment situation of young Hungarian graduates in the engineering fields, a slightly heterogeneous picture can be drawn. According to current graduate career tracking data based on the integration of administrative databases, the highest earners among young graduates are those who are employed in areas of information technology and engineering [NYÜ 14]. Furthermore, one of the highest levels of vertical and horizontal matching can be observed among those with degree in engineering or IT. Although young engineers and IT professionals perform above average in these dimensions (i.e. they are in an advanced position as concerning for their income level, vertical and horizontal matching, their level of skill utilization is just slightly above the average as shown by the results of the Hungarian Career Tracking Survey. Table 7.2 presents the ratio of different graduates' self-reported skill matching. In this comparison, only those working in their studied profession were observed in order to capture the determinants of skill matching more precisely. These primary data show that the Hungarian engineers early in their career are not in a prime position in

1 Source: CEDEFOP 2010, ILO 2013, OECD 2011.

terms of subjective skill utilization, despite their high-level structural congruence. More than a half of them (54.4%) perceived incongruence between their skills and job requirements.

Field of study	Skill mismatch	Skill match	Total
Agriculture	46.0	54.0	100.0
Arts	31.0	69.0	100.0
Economics	63.1	36.9	100.0
Engineering/IT	54.4	45.6	100.0
Humanities	45.0	55.0	100.0
Law	26.7	73.3	100.0
Medical and healthcare	19.2	80.8	100.0
Natural sciences	43.3	56.7	100.0
Public administration and national defense	68.2	31.8	100.0
Social sciences	57.2	42.8	100.0
Sports	30.9	69.1	100.0
Teacher's training	22.3	77.7	100.0
Total	48.4	51.6	100.0

Table 7.2. *Ratio of graduates* with skill matching (in % of graduates in horizontally matching positions) (N = 8,137)*[2]

We can conclude that although in terms of the labor market prospects newly graduated engineers in Hungary are in a favorable situation compared to other professions, and their skills seems to be slightly better harmonized with the job requirements than the Hungarian average, still almost every second person reported a skill mismatch in their current position. This raises serious issues related to engineers' training.

The matching of skill supply and labor market demands is shaped by several individual and institutional factors. If we want to understand the effect of these tendencies on skill matching, there are three different dimensions of the problem on the research agenda:

– Input aspects: the investigation of those factors that shape skill supply, e.g. the skill provision of the education system and skill policies.

2 Edited by the authors. Source: Hungarian Graduate Career Tracking – Institutional survey 2014. Educational Authority. *Full-time students.

– Processes: investigation of how skills are implemented in specific work environments.

– Outcomes: what are the social and economic costs and consequences of skill matching and especially of skill mismatch?

This chapter deals with the input issues and to some extent with those structural characteristics of firms that can be associated with their skill use practices. The core question is what individual and structural factors determine skill matching in the case of newly graduated engineers? Skill matching is determined by several factors that generally occur simultaneously. There are two main institutions representating skill supply and demand: the (higher) education system providing necessary skills and work organizations accounting for the proper utilization of them. There are more possibilities to provide plausible explanations for skill matching or skill mismatch; in our study, the focus is on the following determining factors: (i) individual abilities, attitudes toward learning and efforts in skill acquisition, (ii) skill provision characteristics of higher education, and (iii) skill utilization practices of workplaces.

Individual factors obviously play a decisive role in skill matching. The graduates' personal aspirations, learning attitudes, their motivation to learn and their cognitive capabilities strongly determine both the quantity and quality of skills they possess. Good grades and extra work during their studies (e.g. academic performance), and also their willingness to study abroad, indicate effective skill acquisition and may serve as signals for employers that potential employees will adapt to the job requirements quickly and perform well thereafter [CIP 15]. In addition, academic performance, work experiences during studies may also contribute effectively to skill acquisition and to the development of competencies that support the application of theoretical knowledge in practical situations.

Aside from employees' individual characteristics and capabilities, there are also institutional effects that frame the actors' room for maneuver. In the first place, educational institutions in general and higher education in particular are important determinants of skill matching as they account for the provision of the relevant skill supply. Effective knowledge transfer in higher education is determined by several factors, such as training curricula, academic versus practical orientation of study programs, length of programs (which from the perspective of the individual can also be interpreted as an investment as educational attainment acts as potential signals for employers), innovative teaching/didactical methods, etc.

The other important institutional determinant is the knowledge utilization practices of the companies, which receives less attention in skill matching research than it deserves. Skill utilization refers to the appropriate application of skills and competences in order to maximize performance and to achieve organizational/business goals [PAY 10]. Effective skill utilization is determined by the organizational context that encompasses the internal division of labor, as well as the characteristics of work tasks, like autonomy in work, job complexity, the extent of specialization, the ratio of routine tasks and the need for problem solving. Knowledge utilization is often referred to as "absorptive capacity", that is the "ability to recognize the value of new information, assimilate it and apply it to commercial ends" [COH 90].

As a detailed analysis prepared by CEDEFOP points out, absorptive capacity depends on both human and structural factors [CED 12]. The former refers among others to the level of knowledge, skills and competences (KSC) possessed by the employees and all the efforts to develop them, while the latter is associated with such "soft" and "hard" issues like organizational culture, IT infrastructure and work organizational practices that "provide a range of positive benefits for employees, including the opportunity to use their skills at work" [PAY 10].

Evidence suggests that innovative work organization practices with high absorptive and learning capacities are associated with efficient skill matching and their distribution largely depends on structural factors such as the industrial sector, company size and the composition of the workforce [VAL 09].

7.3. Research question

Alongside their advantageous labor market position, a relatively high level of self-reported skill mismatch can be identified in the case of newly graduated engineers. These statistics, however, should be interpreted carefully. On the one hand, it may reflect the weak correspondence between the skills provided by the higher educational levels of engineers and job requirements; on the other hand, it may also indicate the inadequate utilization of graduates' skills by the companies.

Our primary research question was what the determinants of perceived skill matching are in case of newly graduated engineers? In order to reflect the complexity of the issue, an empirical model has been developed in order to take the various aspects of potential determinants into account. The model tries to capture both individual and structural determinants simultaneously and seeks to answer the following questions. What individual factors affect young engineers' skill matching? What are the impacts higher educational institutions have on skill matching? And finally, what structural characteristics of the workplaces can be associated with skill matching? In section 7.3.1, a more detailed description of the investigated determinants will be provided accompanied by our assumptions concerning their causal effects.

7.3.1. *Effect of academic performance*

Although the effect of social origin both on academic performance and on employability is undeniable, the theory of meritocracy must also be considered when examining labor market success of young graduates. This conception is based on an idea that in the developed industrial societies, instead of demographic, geographic, and social conditions, it is the individual performance that gains an ever increasing role in the achievement of social status [TRE 70, JAC 06]. Consequently, in our view better academic performance leads to more favorable labor market conditions.

7.3.2. *Effect of labor market involvement*

For the purposes of the explanation of graduates' labor market success in terms of skill matching, determinants of the individual strategies also need to be taken into consideration besides the sociodemographic or education macrolevel factors. In the framework of this approach, we presume that an individual's connection to the labor market during his or her studies has a strong effect on his or her early labor market career after graduation. Although there is a large amount of empirical evidence suggesting that employment of students during their studies enhances their future labor market success, studies point out an important distinction: only work experience collected in the field that corresponds to one's prospective profession – as opposed to employment that is not related to one's field of expertise – contributes to successful transition to the labor market. At an international level, on the basis of HEGESCO[3] data, Róbert and Saar [RÓB 12] investigated the effect of employment during one's studies on finding a job after graduation, and concluded that while employment related to the individual's prospective profession exercises positive influence on his or her entry into the labor market, employment unrelated to his or her professional field does not have such favorable effects. The analysis of the Hungarian Graduate Tracking System data shows that for recent graduates profession-related work performed during one's higher education studies has a favorable effect on the success of career and professionalism as factors of success [VER 14]. In this respect, high-level employment rates during studies measured among graduates in engineering and IT programs predict favorable conditions for employability (i.e. for skill matching).

7.3.3. *Effect of educational background*

Educational background is also assumed to affect the future performance of the graduates. On this basis, we separated the interpretation of educational career track into two categories: structural and individual components. With regard to skill utilization, duration and level of studies can be considered as a relevant factor that is represented in BA/MA program distinction in this case. The Bologna system, e.g. the bachelor/master structure, represents a clear hierarchy according to its gradual structure. Several studies of the new system made an attempt to describe and examine the chances of students for BA/MA transitions and its labor market benefits [VER 13, RÓB 13]. On the one hand, the purpose behind the introduction of the linear system was to ensure simplification and consistency in the European Higher Education Area. On the other hand, the significantly reduced time required to acquire qualification through BA training for a large number of students was also an important point [KEH 06]. Consequently, the introduction of bachelor programs boosted the requirements and expectations for the better employability of young graduates. Concerning the implementation of the Bologna system, one of the central arguments was that the two-cycle system would increase the practice orientation of higher education, therefore especially in case of the BA graduates a more harmonized skill matching can be predicted. On the other hand, duration of studies seems to be a relevant predictor of skill matching in the case of highly specialized job requirements,

3 Higher Education as a Generator of Strategic Competences (HEGESCO) project: http://www.hegesco.org.

since a longer period of training can result in more developed skills. This is somewhat contradictory, considering that practice-oriented skill development is strongly linked to BA level programs as a pronounced objective, while BA studies indicate a shorter time spent in education.

For the purposes of explaining the effect of educational background on labor market success, the individual investment effect is also need to be considered. It is based on the economic approach of the theory of human capital that interprets decisions regarding education as an investment that – through the investment of money and time – increases the individual's human capital manifested in knowledge and skills [MIN 74, BEC 64]. The return of investments in education can be interpreted at the level of the individual (as favorable labor market conditions) and at the level of the society (as the improvement of the productivity of the economy). In our research, while interpreting determinants of engineers' early labor market success besides educational performance, the higher level of education attained and longer labor market involvement are also considered as an investment.

7.3.4. *Effect of workplace characteristics*

As referred to earlier skill utilization practices in workplaces also account for the proper matching of skills and job requirements. Skill utilization depends on the structural characteristics of work organization, especially on the internal division of labor and the peculiarities of work tasks. The data set that has been used in our analysis does not contain direct measures of such work organization practices. Therefore, the measurement of the skill utilization effects on skill matching in our models rests on the application of proxies. Since organizational practices supporting extensive skill utilization are strongly influenced by structural factors of workplaces, like size, industrial sector and ownership [VAL 09], the following proxies have been used in the analysis.

– Proxy 1: company size measured by the number of employees; size can be associated with the complexity and internal division of labor and also indicates the complexity of work tasks and the extent of the available resources [MIN 79]. According to our assumptions micro-, small- and medium-sized enterprises (SME) have less developed work organization and knowledge use practices than larger companies. Skill matching in SMEs is predicted to be less prevalent.

– Proxy 2: industrial sector: this is an important indicator of the prevalence of advanced work organization models. Evidence suggests that work organizations in manufacturing are more mature in their technology and manpower use than their counterparts in services [LOR 04]. In the service sector, however, the prevalence of organic, innovative and knowledge-intensive organizations representing dynamically changing skill utilization patterns is more prevalent [SAL 06, VAL 09]. Different sectors are in different business cycles that does not exclusively result in job creation and job destruction (labor demand) but also in dynamically changing skill utilization practices [ARC 12, CED 10].

– Proxy 3: ownership: the Hungarian economy can be characterized by segmented ownership relations. Evidence suggests that different types of ownership (e.g. multinational companies versus domestic ones, state-owned versus private organization) have an impact on the companies' performance as well as on their skill utilization practices [MAR 08]. Ownership indicates the integration of firms into different levels of various value chains, their market conditions and the characteristics of competition they have to cope with, and all these factors effect skill use practices [GER 05].

7.4. Data and methods

The analysis is based on the data set of the Hungarian Graduate Career Tracking System. This regular online graduate survey has been carried out by HE institutions since 2010. The database from 2014 covers the population of Hungarian graduates 1, 3 and 5 years after their graduation and contains the data of up to 22,000 participants. The method of data collection is an online survey, and the average response rate was 12% in the year of the examination. The quality of the data was improved by a cell weighting procedure while the distribution of the sample was modified on the basis of the proportion of the population by gender, year of graduation, study fields and type of study program.

The examination of the research question required the reduction of the core database. For this purpose, we restricted the data to engineering/IT graduates in full-time programs who were employed at the time of the survey and who worked in study-related jobs (i.e. in horizontally matching positions). The resulting database consists of 1,850 observations. During analysis, IT and engineering study fields were merged due to the specificity of the Hungarian classification system in order to increase comparability with the international ISCED fields of study classification. According to this, in our investigation the engineering study field covers computing, engineering, manufacturing and construction ISCED categories (coded by 48, 52, 54 and 58 in ISCED classification).

The analytical method we applied is a binary logistic regression model where the odds ratio for skill match among young engineers (as dependent variable) was predicted. The values of the skill matching level as our dependent variable are based on the self-assessment of the respondents of the online survey on a 5-point Likert scale[4]. The initial content was compressed into a dummy variable, in which 1–3 values were merged into the skill mismatch category and the remaining 4–5 values were interpreted as skill matching. For the purpose of the explanation of skill matching among young engineers, five consecutive subgroups of independent variables were built into the binary logistic regression model, namely "performance" and "involvement" as individual effects and "educational background" and "workplace characteristic" as structural effects were included. In the fifth step, several sociodemographic characteristics were built as controls.

The blocks of independent variables were as follows:

– Block 1: Performance (individual effects):

  - subjective academic performance above average (binary variable).

– Block 2: Involvement (individual effects):

  - non-professional work during studies (binary variable);

  - professional work during studies (binary variable);

  - international student mobility involvement (binary variable).

– Block 3: Higher educational background (structural effects):

  - bachelor degree (binary variable);

  - graduation year 2013 (binary variable).

4 The relevant question was: *To what extent do you use in your current work the knowledge and skills acquired during the studies that constitute the basis of the questionnaire?* [1] Not at all; [2] To a slight extent; [3] To a medium extent; [4] To a great extent; [5] Completely.

– Block 4: Workplace characteristic (individual/structural effects):

- company size: micro (up to 9 employees) (reference category); medium (50–249 employees); large (above 250 employees);

- ownership: foreign-owned operating abroad; foreign-owned operating in Hungary; Hungarian owned (reference category);

- sector: industrial; services (reference category) (binary variable).

– Block 5: Sociodemographic background (contextual effects):

- male (binary variable);

- type of residence at 14: rural; village (reference category) (binary variable);

- academic background (binary variable);

- favorable economic background at 14 (binary variable).

In the following, we provide a short description of the independent variable sets built in the model. Academic performance during studies is measured by the relative educational achievement compared to other students. This solution aims to eliminate the differences between the overall evaluation levels of various faculties and institutions. Graduates who judged that their performance during studies was better than other students on the same training were separated into "subjective academic performance above average" category (value 1). The remaining cases were coded to 0. In the case of "Bachelor degree" dummy, the master and one-cycle level programs were merged and coded to 0. With regard to the graduation year dummy variable, year 2013 – i.e. the most recent graduates – was separated (as 1) from the group of graduates of 2011 or 2009 (0). Regarding workplace, characteristic companies were divided into three categories based on the number of employees from which small size companies represent the reference category in the model. During the categorization of ownership, three types were developed based on localization of operation and owner's national identity. In this case, Hungarian-owned firms serve as reference category. Our last variable describing workplace characteristics separates firms into economic sectors from which the industrial sector was highlighted (value 1) in the dummy. Among sociodemographic controls, the academic background dummy includes graduates with at least one parent with a degree against those who have no such background. The favorable economic background at 14 variables is based on the respondents' self-declaration for their families' financial situation in relation to their environment.

7.5. Empirical findings

In Table 7.2, we summarize the results of our explanatory model presented above. A binary logistic regression model was built in order to isolate the impact of several individual and structural factors on skill matching among young engineers. The interpretation is based on the values of odds ratio presenting the independent effect of each background variable on the chance of obtaining a high-level skill match in work. We assumed that better educational performance predicts a more favorable skill match. To examine the signs of merit-based selection, the relative (compared to other students of the same program) academic performance of graduates was used. The model proved the significant positive effect of academic excellence. With regard to additional educational investments, gaining professional work experience during studies significantly raises the odds ratio of skill matching after graduating in the engineering field as was expected. Surprisingly, non-professional work during studies has a negative impact on skill matching. It seems that participation in international student mobility is not significant for future skill match. In the case of the higher educational background regarding the training level or the time of graduation, we did not find a significant effect on the chance of a skill match. The effects of workplace characteristics examined as the fourth dimension corresponded to the expectations. As for company size the odds ratio of skill matching was significantly lower among employers of medium-sized and large companies than in case of their small-sized counterparts chosen as a reference group. In addition, ownership also seems to be an important factor influencing skill matching. Working at foreign-owned companies operating in Hungary decreases the chance of skill matching compared to both foreign-owned companies operating in Hungary and companies in Hungarian ownership. In addition, our empirical results confirmed our assumptions that the industrial sector is an important determinant of skill matching. Firms operating in manufacturing offer a more favorable work environment in terms of skill matching to newly graduated engineers than companies in the service sector.

Blocks	Background variables	Odds ratio (Exp B)
Block 1: Performance	Subjective academic performance above average	1.491***
Block 2: Involvement	Professional work during studies	1.214+
	Non-professional work during studies	0.693**
	International student mobility involvement	1.201
Block 3: Higher educational background	Graduation year 2013	1.192
	Bachelor degree	0.941
Block 4: Workplace characteristic	Company size: Medium (reference small)	0.754+
	Company size: Large (reference small)	0.636**
	Ownership: foreign-owned operating abroad (reference Hungarian owned)	1.571*
	Ownership: foreign-owned operating in Hungary (reference Hungarian owned)	0.776+
	Industrial sector (reference service)	1.387**
Block 5: Socio-demographic background	Male	1.067
	Rural background	0.904
	Academic background	1.010
	Favorable economic background at 14	1.158
Model statistics	HL test significance	0.811
	(Pseudo) R Square (Nagelkerke)	0.072
	Intercept	0.912

Table 7.3. *Determinants of skill matching of newly graduated engineer and IT specialists*: regression results (binary logistic regression model for skill matching among young engineers) Exp(B)*[5]

7.6. Discussion

The aim of this chapter is to analyze individual and structural determinants of skill matching among young Hungarian engineers. For a deeper understanding of the problem, we provided a detailed description of the skill matching phenomenon and its determinants aiming to explore why despite the relative favorable employment status of young Hungarian engineers both in terms of soft and hard labor market indicators, regarding skill utilization their situation seems less favorable.

5 Source: Graduate survey 2014 – Educational Authority, N =1,850, Significance: ***$P < 0.001$; **$P < 0.01$; *$P < 0.05$; +$P < 0.1$; *working in horizontally matching positions.

In order to explain this situation, we examined the effect of several background determinants on skill matching, considering the significance of individual characteristics (i.e. academic performance and former labor market involvement) and structural frameworks (i.e. educational and workplace characteristics). During this, our assumptions were built into an explanatory model on the distribution of skill match among young specialists in engineering/IT fields.

Assumptions condensed into research questions were confirmed by the empirical analysis in relation to the academic performance, the labor market involvement and the structural characteristics of the workplace. All these background determinants proved to be relevant regarding the skill utilization of young engineers in the Hungarian context. Their effect remained significant even if it was controlled by the sociodemographic background of graduates. However, regarding the educational background, attaining a higher level education (to undertake a more investment) did not necessarily increase the chances of skill match significantly. The effect of high-level academic performance on skill matching can be explained by two factors. It may serve as signals for the employers to employ graduates with better capabilities and, as it can be associated with a high level of skills as well, they may be employed in better jobs that require intensive use of high and specialized skills. Concerning work experience during studies, it is somewhat surprising that non-professional work negatively correlates with skill matching. A possible explanation could be that non-professional work has a deteriorating effect on skill acquisition by taking place at the expense of the time that should be devoted to studies or collecting professionally relevant experience.

As for the structural characteristics of workplaces, our results only partially confirmed our hypothesis. We assumed that skill matching is more problematic in the case of the SMEs than in large companies, but our results indicate a reverse relationship between size and skill match. According to our potential explanation, the smaller a company is, the less developed the internal division of labor is. This means that the work task is not specialized, and the scope of activities in a job may be wider than in the case of larger companies. Paradoxically, it also may imply that control is formalized and there is a need for greater autonomy at work. Complex jobs, rich and diverse work tasks contribute to preserving high motivation and favor effective skills and manpower use. For the ownership relations, our results are quite surprising. Corresponding to our original assumption, skills and job requirement seem to fit best in the case of foreign-owned companies, but there is a difference between foreign-owned companies operating in and outside Hungary. It is far from an easy task to provide a satisfactory explanation for these findings. It can be argued that firms operating on more advanced markets are in a better position in the international value chains with respect to producing high value added products and services. To be able to upgrade in value chains, effective exploitation of skills is an important requirement and companies exposed to more intensive, global competition are pressed more to apply work organization practices that favor effective skill use. Corresponding to our original assumptions, companies in the manufacturing sector are in a better position when harmonizing skills and job requirements than their counterparts in services. A possible explanation can be that manufacturing is a different "maturity cycle", which means companies are better adapted to the well-known technology they use and to its changes. It indicates that in their case, because of the larger stock knowledge and experience having been accumulated in the last centuries, technology, work organization and manpower use practices that are well established and harmonized and the institutional environment (e.g. education other institutions of knowledge transfer and creation) support them in better exploiting their technological abilities and relevant skills (that are more standardized than in case of services and related new technologies).

7.7. Conclusions

In order to address the gaps in our knowledge concerning skill match in the case of newly graduated engineers in our contribution, we aim at providing statistical evidences to explain the role of various determinants of the phenomenon. The novelty of our approach rests in the combination of individual and institutional factors in one explanatory model. Our main findings call to attention that research in the field should be extended to the analysis of possible consequences of skill matching and skill mismatch for job quality and job satisfaction, and more attention should be devoted to the mechanisms of skill utilization at the company level, as well. As mentioned previously, skill deficits may lead to unfavorable consequences in competitiveness and economic growth, underutilization of skills entails serious losses of individual and social resources.

In the Hungarian context, governmental initiatives have been made in order to bridge the gap between the knowledge provision of the higher education institutions and the labor market requirements. In this framework, German-type dual programs have been introduced in five study fields, including engineering/IT. The dual programs combine academic studies provided by the participating university and work-based learning with a company most often from a private sector. Our results, however, call attention to the fact that policies aiming at reducing skill mismatch on the labor market should focus both on skill provider institutions and skill utilizers simultaneously. Our findings suggest that there are structural differences between companies with respect to their skill use practices, therefore in order to tackle the problem, besides advanced educational policies, those work organization methods should be promoted that ensure greater autonomy and task discretion, and make skill utilization more profitable for both employers and employees.

7.8. References

[ARC 12] ARCHIBUGI D., FILIPPETTI A., FRENZ M., The impact of the economic crisis on innovation: evidence from Europe, CIMR Research Working Paper Series, Working Paper No. 5, Birbeck University of London, 2012.

[BEC 64] BECKER G. S., *Human Capital: A Theoretical and Empirical Analysis, with Special Reference to Education*, University of Chicago Press, Chicago, 1964.

[CED 10] CEDEFOP, *The Skill Matching Challenge. Analysing Skill Mismatch and Policy Implications*, Publications Office of the European Union, Luxembourg, 2010.

[CED 12] CEDEFOP, "Learning and innovation in enterprises", Research Paper No. 27, Publications Office of the European Union, Luxembourg, 2012.

[CIP 15] CIPD, Over-qualification and skills mismatch in the graduate labour market, Policy report 2015,Chartered Institute of Personnel and Development, London, 2015.

[CLA 10] CLARK R., ANDREWS J., "Researching primary engineering education: UK perspectives, an exploratory study", *European Journal of Engineering Education*, vol. 35, no. 5, pp. 585–595, 2010.

[COH 90] COHEN W.M, LEVINTHAL D.A, "Absorptive capacity: a new perspective on learning and innovation", *Administrative Science Quarterly*, vol. 35, pp. 123–138, 1990.

[GER 05] GEREFFI G., HUMPHREY J., STURGEON T., "The governance of global value chains", *Review of International Political Economy*, vol. 12, no. 1, 2005.

[ILO 13] ILO, Global Employment Trends for Youth 2013: *A Generation at Risk*, International Labour Office, Geneva, 2013.

[JAC 06] JACKSON M., Meritocracy, *Equalsoc Research Group Education, Social Mobility and Social Cohesion (EDUC)*, pp. 58–83, December 2006.

[KEH 06] KEHM B., TEICHLER, U., "Which direction for bachelor and master programmes? A stocktaking of the Bologna process", *Tertiary Education and Management*, vol. 12, pp. 269–282, 2006.

[LOR 04] LORENZ E., MICHIE J., WILKINSON F., "HRM Complementarities and Innovative Performance in French and British Industry", in CHRISTENSEN J.L., LUNDVALL B.A. (eds), *Product Innovation, Interactive Learning and Economic Performance*, Elsevier, Amsterdam, 2004.

[MAR 08] MARTIN R., "Post-socialist segmented capitalism: the case of Hungary. Developing business systems theory", *Human Relations,* vol. 61, no. 1, pp. 131–159, 2008.

[MIN 74] MINCER J., *School, Experience and Earnings*, National Bureau of Economic Research, Inc., New York, 1974.

[MIN 79] MINTZBERG H., *The Structuring of Organization*, Prentice Hall, Englewood Cliffs, 1979.

[NYÜ 14] NYÜSTI S., VEROSZTA, ZS., Hungarian Graduate Career Tracking 2014 - Integration of Administrative Databases. Preliminary report, Education Non-Profit LLC, Budapest, 2014.

[OEC 11] OECD, *Employment Outlook 2011,* OECD Publishing, 2011,

[PAY 10] PAYNE J., "In search of a measurement and evaluation framework for skill utilisation", in PAYNE J. (ed.), *Skill Utilisation: Towards a Measurement and Evaluation Framework*, SKOPE Cardiff University, Cardiff, 2010.

[RÓB 12] RÓBERT P., SAAR E., "Learning and working: the impact of the 'double status position' on the labour market entry process of graduates in CEE countries", *European Sociological Review*, vol. 28, no.6, pp. 742–754, 2012.

[RÓB 13] RÓBERT P., VEROSZTA ZS., "Even with a degree? Hard decisions and strong selections in the transition from university to the labor market among Hungarian graduates during the crisis, 2007-2011", *21st Annual Workshop of the European Research Network on Transitions in Youth School-to-Work,* Berlin, Germany, September 11–14, 2013.

[SAL 06] SALTER A., TETHER B., *Innovation in Services (Through the Looking Glass of Innovation Studies)*, Imperial College London, London, 2006.

[TRE 70] TREIMAN D., "Industrialization and Social Stratification", in LAUMANN O.E., (ed.), *Social Stratification: Research and Theory for the 1970's*, The Bobbs-Merrill Company, Indianapolis, 1970.

[VAL 09] VALEYRE A., LORENZ E., MAKÓ C. *et al.*, Working conditions in the European Union: work organisation, Report, 2009.

[VER 13] VEROSZTA Z., "The way to master programmes – an examination of the selection mechanisms in the bachelor/master transition in higher education", in GARAI O., VEROSZTA Z. (eds), *Hungarian Graduates 2011,* Educatio Társadalmi Szolgáltató Nonprofit Kft., Budapest, pp. 9–37, 2013.

[VER 14] VEROSZTA Z., "Employment effects on planned higher educational career track", in KISS L. (ed.), *The Social Dimension of Higher Education: The Results of Eurostudent V in Hungary*, Educatio Társadalmi Szolgáltató Nonprofit Kft., Budapest, pp. 53–67, 2014.

PART 3

Pedagogies of Innovation

8

Swimming with Sharks without Being Eaten: How Engineering Students can Learn Creativity, Entrepreneurial Thinking and Innovation

"No matter who you are, the creative spirit can enter your life. It is at hand for anyone who has the urge to tinker, to explore new possibilities, to leave things a little better than before" [GOL 92].

"Technological innovations occur when a need arises or an opportunity presents itself" [NAT 04].

"In modern society, engineers are increasingly expected to move to positions of leadership and to take on additional roles as entrepreneurs" [CRA 14].

"Preparation for entrepreneurship, that is, the starting of a new company, involves unique competencies that can be learned" [AUL 13].

8.1. Introduction

How can engineering students learn to boost their creativity and to think like successful entrepreneurs aiming at innovativeness? Since the

Chapter written by Claudius TERKOWSKY, Tobias HAERTEL, Anna-Lena ROSE, Liudvika LEISYTE and Dominik MAY.

discussions on the quality and importance of higher engineering education facing vast challenges of a globalizing world have gained momentum, three subtopics are coming more and more to the fore of higher engineering education:

– *fostering students' creativity* in order to strengthen future engineers' abilities to generate ideas, to find optimal solutions for complex problems and to invent new processes and products;

– *nurturing students' entrepreneurial skills* in order to transform these novel processes and products into business ideas;

– *encouraging students' potential to innovate* in order to exploit these processes and products as value propositions in economic and social spheres [CRA 14, HAE 12, HAE 16a, MAY 15].

Only if creative ideas become innovations, are they able to create impact and be of use. Or as Goleman, Kaufman and Ray put it: "When you take a new approach to what you're doing–and the new approach works–you're using your creativity. When you go beyond the traditional ways of solving a problem with a success that influences others, your creativity takes on a vital social dimension" [GOL 92].

Focusing on defining the concept of innovation, Edison, Ali and Torkar see "innovation is an iterative process. In this process, idea generation or adoption, development of products (by using competencies inside and outside the organization) and introduction of products to marketplace and implementation of processes in organisation are the major activities" [EDI 13]. The authors differentiate two aspects of innovation that are paramount:

– the *degree of novelty* (new to the company, new to the market, new to the industry, new to the world);

– the *degree of impact* (incremental innovation, market breakthrough, technological breakthrough, radical innovation).

Consequently, Christensen and Overdorf underscore that disruptive innovation (defined as innovation with a destructive impact on existing value chains) is key to prospective achievement in economy and business [CHR 00].

Discussing the connection between entrepreneurship and innovation, Drucker sees innovation as a core activity of entrepreneurship: "Innovation is the specific function of entrepreneurship, whether in an existing business, a public service institution, or a new venture started by a lone individual in the family kitchen. It is the means by which the entrepreneur either creates new wealth-producing resources or endows existing resources with enhanced potential for creating wealth" [DRU 02].

Shifting the perspective on the individual capacity of entrepreneurs, Byers *et al.* define entrepreneurship as "a personal skill and motivation that draws a person to engage his abilities and efforts in the creation of new products and services". For them, personal skills in this relation are self-awareness, accountability, emotional coping, teamwork, creativity and risk taking [BYE 13]. On top of that, according to May, it is the "courage to create" [MAY 94] that builds the basic core of creativity. The newer, the more original, deviant or remote the idea, the more courage is necessary to share and shield it, to pursue it against the risk of refusal. Bertram and Preißing assert that being creative is at the same level with "navigating in open systems", which is inherent to all creative actions. Being creative means to inevitably leave the "closed system" for a while: the safety-mediating routines, traditions and norms. The creators have to navigate in the open system without knowing which response they might get (positive or negative) for their creative actions – but they had better expect one [BER 07].

As a provisional summary it can be stated that the ability to be creative is fundamental and prerequisite for successful entrepreneurial thinking and for imagining intended innovative impact. It is the actual core process of innovativeness. Finally, while creativity is freely allocated to divergent thinking for broad idea generation, entrepreneurship is more assigned to convergent thinking ending up in a concrete business proposal aiming at desired innovation. Moreover, in ordinary higher engineering education contexts, the expression of creativity is clearly viable, it can be instructed and observed while the desired impact of innovation is more or less an imagination of the unpredictable future.[1] This raises three essential and challenging questions for higher engineering education:

1 At least the German model of higher engineering education is not exactly well known for rewarding students' abilities to create and innovate [LEM 17, MOR 17].

1) How can creativity, entrepreneurship and innovativeness be fostered and nurtured in higher engineering education?

2) How can students learn to apply creativity techniques and business development frameworks to solve engineering problems, and by these means, foster and nurture their entrepreneurial thinking skills and their ability to create and innovate?

3) How can students nurture their agency to create and develop the courage to take risks to really "thinking big" and "out of the box" for radical innovation – above all in risk-free and benevolent pedagogical settings?

The collaborative research project "ELLI – Excellence in Teaching and Learning in Engineering Science"[2] funded by the German Federal Ministry of Education and Research focuses on these challenges with a large work package aiming at fostering creativity, entrepreneurial thinking and innovativeness in higher engineering education. Within this project, a special tutorial concept at TU Dortmund University has been developed and adapted in order to accompany a sequel of regular entrepreneurship and project management seminars, which are open for students from the faculties of engineering and business and economics with practical work. These "Shark Tank Experience" tutorials [HAE 16b], taking place every semester, aim at fostering engineering students' creative potential, entrepreneurial thinking and innovativeness for developing business ideas. The Shark Tank Experience is a pedagogical simulation adapted from the American reality TV show Shark Tank[3] in which entrepreneur candidates apply for venture capital from investors [DEH 12, WIL 12]. The TV show Shark Tank mainly

2 ELLI is a large German collaborative research and development project on higher engineering education carried out by partners from RWTH Aachen University, Ruhr-University Bochum, and TU Dortmund University. Part 1 ("With creativity, mobility and interdisciplinarity to Bologna 2.0", funding: €11.1 million) took part between 2011 and 2016; part 2 ("In times of increasing virtuality, globalization and interdisciplinarity to industry 4.0", funding: €10.5 million) proceeds from 2016 to 2020.

3 The American TV show Shark Tank initially performed in 2009 and is a derivative of the international format Dragons' Den, which dates back to the Japanese TV Tigers of Money originated in 2001. Franchise versions of the show have been produced in almost 30 countries around the world (not in France, yet). Shark Tank shows entrepreneur candidates presenting a business idea to a board of investors ("sharks") in order to gain venture capital. The investors examine the business idea after the candidate has presented the pitch. This will lead either to a viable business proposition of an investor interested in investing capital in return for equity, or lead to refusing the investment due to uncovering of serious weak points in the business idea [DEH 12, WIL 12].

deals with presenting and judging pitches of business proposals. This approach has been transformed into a role play scenario, in which students have to generate business ideas for successful start-up enterprises. To this end, they have to prepare, present and defend their pitch in front of an audience playing the roles of tough and cherry-picking venture capitalists, which probe the ideas and finally decide to invest or not to invest.

Furthermore, this experiential essay on the case study Shark Tank Experience introduces some basic considerations regarding entrepreneurship training in higher engineering education. It is followed by a very brief summary of the preferred and therefore considered contemporary concepts of creativity in higher engineering education, the deploying of creativity techniques and of practicing the courage to create. Afterward, it introduces the case of the Shark Tank Experience, including intended learning objectives, learning activities and grading considerations and discusses deployed methods, followed by a presentation and discussion of some results of running the scenario. After this, a short section about expanding the concept in planned future work will be presented. Finally, it ends with some concluding and probably slightly provoking thoughts on fostering creativity, entrepreneurship and innovativeness in higher engineering education.

8.2. Basic considerations regarding entrepreneurship and creativity

8.2.1. *Entrepreneurship in higher engineering education*

According to Lo, entrepreneurship education in general can be described and defined as the means and approaches used to teach students to start novel businesses and understand how to run such businesses [LO 11]. To this end, students in engineering entrepreneurship programs are trained to develop team work, effective communication, independent thinking, open-ended problem solving, design for end users and understanding business basics skills [BYE 13]. These skills can be gained in a foreseeable amount of time. However, in most engineering education programs, these novel topics ended up as add-ons to train entrepreneurship as an additional competence without appropriate integration into engineering and sciences subjects [DOW 05, GOL 06]. In recent times, engineering schools have responded to these demands with fairly reformed policies: by fostering entrepreneurship skills through management and business courses especially for engineers [ASM 08] or by conceptualizing entrepreneurship as a critique and the capacity to provide significant problem solving for a company facing a series of new challenges [JØR 11]. Nevertheless, the inclusion of creativity, entrepreneurship and innovativeness in higher engineering education curricula calls for a transformation of mind sets and personality on the part of teaching staff to contribute to – or at least tolerate – adjustments in the engineering syllabus [BYE 13].

8.2.2. *Contemporary concepts of engineering creativity*

Summarizing a lot of relevant work in this field, Plucker, Beghetto and Dow define that creativity is "the interaction among aptitude, process and environment by which an individual or group produces a perceptible product that is both novel and useful as defined within a social context" [PLU 04]. Cropley and Cropley transfer important results of the broader discussion about creativity into the context of engineering [CRO 05]. While creativity in general mainly deals with generating ideas and novel solutions to problems, engineering creativity delves into bringing forth technological solutions to the task at hand, e.g. by designing and developing artefacts, processes, models, systems and services [CRO 16]. Cropley and Cropley coined the term "functional creativity" to distinguish novel, useful creativity from that which is purely aesthetic [CRO 10]. Finally, based on their comprehensive empirical research, Haertel, Jahnke, Terkowsky, Wildt and others conceptualize their Dortmund model of six different facets for fostering creativity in higher (engineering) education (The "6F-model") (see Figure 8.1). According to the latest revision, these six facets are:

1) reflective learning (F1);

2) autonomy: independent self-determined learning (F2);

3) curiosity and other intrinsic motivation (F3);

4) making: learning by creating (F4);

5) multiperspectives (F5);

6) imagining new ideas (F6).

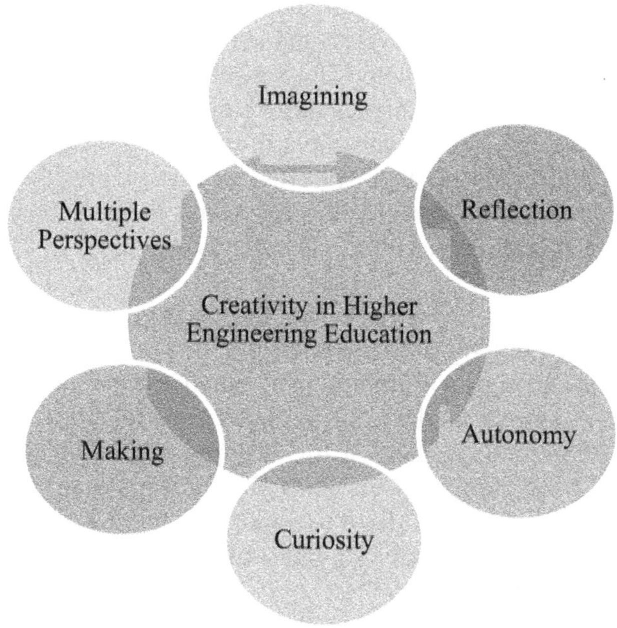

Figure 8.1. *The six facets of fostering creativity in higher engineering education*

From the teacher perspective, student creativity can be categorized according to the 6F, in which teachers "see" students' expression of creativity. The results can provide a new understanding of student creativity from university teachers' perspective that can be useful for reorganizing course designs [HAE 12, HAE 16a, HAE 17a, JAH 17, TER 13, TER 16]. The 6F should rather not be used as a strict framework to evaluate students' creativity, but rather be deployed as a basic guiding perspective for developing educational settings to foster students' creativity.

8.2.3. *Deploying creativity techniques*

Creativity techniques are methods to inspire creative thinking and creative actions. They focus on a variety of factors of creativity, including methods for idea generation and divergent thinking, techniques for creative problem solving and reframing problems, improvisation, motivational aspects, and on the affective influencing of individuals and within groups [BEL 16, BUL 95, CRO 01]. Jackson stresses the effect of the setting on personal creativity. According to him "personal creativity emerges from the ecologies we create to learn, develop and achieve something we value. (...) Learners created their own ecologies for self-determined, self-motivated and self-regulated learning to achieve something they wanted to achieve" [JAC 16]. Investigating the effectiveness of deploying creativity techniques in education, studies reveal that students' creativity profits from getting suitable and appropriate creativity strategies and more elaborate guidance regardless of the academic field [FIN 10, NIU 09].

In the "Shark Tank Experience" tutorial, generating *business ideas* is linked to divergent thinking processes (being aware of remote associations and unusual links, coming to surprising solutions) and personal aspects, such as openness, willingness to take risks, ambiguity and tolerance of uncertainty [CRO 11]. To this end, students learn to apply several reliable creativity techniques like Four Thinking Hats[4], 6–3–5, brain storming or brain writing to stimulate "out of the box" thinking and the emergence of incremental improvements or radical ideas for solving real-world problems in the context of engineering.

Generating *business ideas* in the Shark Tank Experience simulation is linked to convergent thinking processes. These convergent thinking processes are guided with the Business Model Canvas [OST 10], a creativity technique for refining a basic business model according to nine elements of business case development and its inherent relations. These nine elements are *customer segments* (1), *value propositions* (2), *channels* (3), *customer relations* (4), *revenue streams* (5), *key resources* (6), *key activities* (7), *key partnerships* (8) and *cost structure* (9). Instead of investing weeks in writing a classic business plan of about 30 pages, just a pinboard or large piece of paper divided into these relevant fields is necessary. Gradually, the students have to fill in these nine fields with flexibly movable sticky notes, representing their ideas for each of the nine issues. By using sticky notes, the method particularly intends making changes easily and flexibly during the process of generating the business idea.

---

4 The Four Thinking Hats is a complexity reduction of De Bono's Six Thinking Hats made by the project team to give to the students' easier role making and role taking opportunities while applying the method. Moreover: German carnival billycocks used for this are only available in four colors, at the moment. Needless to say, writing and refining the justification for a higher education institution's finance administration until getting the bill paid for 40 carnival melon hats can become a demanding business.

8.2.4. *Unleashing the courage to create by practicing breaching experiments*

While some of the presented cognitive aspects of creativity, entrepreneurship and innovativeness can be relatively easily transformed into *intended cognitive learning outcomes* – and therefore can be learned in the short term like divergent and convergent thinking with the help of creativity techniques – other aspects more related to *intended affective learning outcomes* unfortunately may not [HAE 16a]. Aspects more related to personality change like risk taking, navigating in open systems or displaying a rule braking creative attitude may not be so easy to learn but would need permanent novel experiences and ongoing insight. According to Brodin and Frick, courage is an important part of our self-confidence, and encouragement is a key part of a successful development [BRO 11]. However, Goffman states that the most common rule in all social situations is for the individual to "fit in" [GOF 10]. Breaching experiments intend to disregard these common rules. They involve the conscious violation of social norms in everyday social situations by means of unexpected behavior, especially by breaking unstated but universally accepted rules [GAR 96]. For example, Babbie gave his students the task to fix problems in public places for which they had no regular work contract or order, such as picking up garbage from the street or mending street signs. By observing their own feelings as well as the emotional reactions of random observers, the students had to carve out how these internalized common rules control a social situation and how violating these rules can irritate the interaction between participants of a social situation or staffing of public authorities [BAB 15]. But they also can learn that they usually survive these potentially inconvenient encounters unscathed [HOF 12].

8.3. The Shark Tank Experience tutorial

Currently, the Shark Tank Experience tutorials are integrated parts of two course concepts at TU Dortmund University in Germany, one in the faculty of business and economics, the other in the faculty of mechanical engineering, both of which are open for engineering students.

Since 2015, the faculty of business and economics at TU Dortmund University in Germany has offered regular entrepreneurship seminars that are part of various regular master study programs in the context of innovation management. However, according to the requirements of their course of studies, engineering students can sign in and attend the entrepreneurship seminars as well. While the summer term seminar focuses on “Foundations of Systematic and Strategic Entrepreneurship in Complex Systems – High tech entrepreneurship in International Perspective”, the winter term seminar deals with acquiring and discussing “Academic Entrepreneurship and Applying Academic Entrepreneurship”. Now for the fifth time, the “Shark Tank Experience” tutorials extend these seminars aiming to offer a practical part.

During the summer term 2017, the Shark Tank Experience tutorial has also extended the didactic scenario of a project management course, taking place at the engineering education department of the faculty of mechanical engineering to improve students’ creativity outcomes. During the course, which aims at simulating the work of a development department in a company, the students shall develop a business idea not only as a concept, but also as a real physical prototype. To this end, the students are expected to improve a learn-to-walk toy car by inventing innovative enhancements.[5] Finally, the students are expected to present their prototype during a pitch session to the fictional management, which ultimately decides to go for it or not.

The tutorials have now been carried out for the sixth time since summer term 2015. In the meantime, 88 students have taken part (see Table 8.1). Forty-nine students, about half, were from engineering, while all other students were master students from economics and business administration. The tutorials take place all day on two consecutive days. The involved teachers, trainers and researchers come from the TU Dortmund University’s Center for Higher Education (1), its special Engineering Education Research Group (2), from the faculty of business and management (3) and from the faculty of mechanical engineering (4). All actors form together a multidisciplinary group that is highly unusual to the German engineering higher education system.[6]

---

5 In order to do this, the students can make use of all available machines, equipment and materials in the workshops and labs of the engineering education department at the faculty of mechanical engineering. Additionally, they get a small budget for making reasoned external purchases.

6 The German higher engineering education system is typically characterized by a strong scientific orientation in basic teaching and is rather not known for the promotion of creativity, entrepreneurship and innovation, apart from some attempts with the CDIO approach. CDIO is based on the presupposition that engineering graduates should be able to: “conceive, design,

Term	Total number of students	Engineering students	Number of students in each group
ST 15	12	1	4–4–4
WT 15/16	8	1	2–2–2–2
ST 16	13	3	4–4–5
WT 16/17	8	1	2–2–2–2
ST 17 (1)	10	6	3–3–4
ST 17 (2)	37	37	5–5–5–5–5–4–4–4
Total	88	49	25 groups

Table 8.1.

8.3.1. *Intended learning objectives and learning activities*

The learning activities of the practical tutorials focus on the following intended learning outcomes (ILOs). After the course, students should be able:

1) *to reflect* on their individual set of skills and setting of creativity as well as on conditions that hinder and that foster the expression of their creativity (addressing F1, F2 and F3 of the 6F);

2) *to apply* reliable creativity techniques for idea generation in order to develop business ideas for solving self-defined real-world problems in the context of engineering (addressing F4, F5 and F6);

3) *to prepare, to present* and *to defend* a convincing pitch for the pedagogical Shark Tank Experience simulation (addressing F1, F3 and F4);

---

implement, operate complex value-added engineering systems in a modern team-based engineering environment to create systems and products" [CDI 17].

4) *to practice* breaching experiments ("do something unusual" task) for facing individual risk of refusal-thresholds and reflecting on pushing its limits (addressing F2 and F5).

While ILOs 1–3 aim at practical transformation of academic knowledge and are clearly attainable on a basic level in a tutorial, ILO 4 for fostering risk-taking abilities is more or less a long-term learning objective drawing on personal change by reflection and introspection, which cannot be fully achieved within 2 days. However, students may conclude that there is further need for personal development, if they want to improve going for their own ideas in the long run.

8.3.1.1. *Preparing a pitch by using creativity techniques*

Due to the special workshop character, the students carried out almost all tasks as group work organized in several creative rounds. Initially, there were partial sequences of individual idea generation to prepare subsequent group work. After presenting and clustering all individual ideas, the participants were asked to build groups around the clusters by themselves according to their personal interests. Afterwards, the task for each group was to develop the assigned cluster of ideas into their own business concept and finally to arrange a pitch. Each creative round comprised four stages:

1) *divergent thinking* phase (30-min): students deployed a variety of creativity techniques like classic brainstorming, Four Thinking Heads, force fit with pictures, brain writing, 6–3–5 to devise a first business concept and generating ideas for a successful pitch;

2) *convergent thinking* phase (30-min): students had to decide for the most promising ideas, then specify, concretize and prepare the pitch by applying the Business Model Canvas or related methods to extract a business case;

3) *pitch* phase (10-min presentation, 15-min discussion);

4) *reflection* phase: students had to reflect on and discuss the pitch regarding factors of success or failure and the utilized creativity technique concerning appropriateness, limitations, advantages and disadvantages.

8.3.1.2. *Breaching experiment: "do something unusual"*

In order to perform a breaching experiment, the students had to "do something unusual" between the first and the second day. They had to expose themselves to a self-imposed inconvenient situation and had to reflect on their feelings before, while and after doing it. Moreover, they had to find individual situations and solutions and thus were not permitted to work together in teams [HAE 16c]. While confronting all students with the same unusual-task would lead to a comparability of mastering the inconvenient situation, it would also contain the danger of demanding too little or – even worse – too much of individual participants. Finally, the students had to debate the task and their experiences at the opening of the second day's session.[7]

8.3.2. *Grading considerations*

Since "the essence of creativity is prospective rather than retrospective" [TSA 13] as it is for learning, the course facilitators share serious concerns regarding the current fashion to measure and to assess creativity. Instead, the course facilitators like more to observe curiously and attentively students' creative performances in lieu of rigorous evaluation against questionable criteria. From a pedagogical point of view, if you want to encourage younger people to decide for more risk-taking while creating and innovating, it may be counterproductive to reward them with a blow of an axe in an initial situation.[8] Instead, it might be more educational to give hints for improvements. For this reason, the course facilitators explained to the students that they would receive a grade for their participation at the tutorial, amounting to a fifth of the seminar's overall grading. However, neither the quantity nor the quality of their ideas was graded, but their willingness to participate and their openness for the methods and tasks. At the end of each tutorial during a final evaluating feedback round, students got the

7 For more details, see [HAE 16a, HAE 16c].

8 The course facilitators rather prefer a cautious position like stated in "the three laws" of the science fiction author and inventor Arthur C. Clarke:

1) When a distinguished but elderly scientist states that something is possible, he is almost certainly right. When he states that something is impossible, he is very probably wrong.

2) The only way of discovering the limits of the possible is to venture a little way past them into the impossible.

3) Any sufficiently advanced technology is indistinguishable from magic.

In the course of the centuries, quite a lot of famous examples in human history have piled up, in which the distinguished scientific establishment failed by prophesying that others' ideas remain magical impossibilities [CLA 00].

opportunity to share their learning experiences and to explain what they liked and what they would improve in the tutorial.

8.4. Data collection, data analysis and methods reflection

The main objective of this study is to formatively evaluate the appropriateness of the deployed methods for fostering students' creativity, entrepreneurial thinking, imagined innovation and risk-taking abilities. To this end, the characteristics of students' ideas give valuable feedback to the course facilitators for improving the setting of the tutorials (and not for grading). Due to the special character of the tutorial as a workshop dealing with stimulating and fostering of students' spontaneity and improvisation, a loose qualitative approach for evaluating the Shark Tank Experience has been composed. To this end, data were collected by the following tools:

1) *participant observation* and taking notes by teachers during the tutorials of social and sociotechnical interaction within the tutorials;

2) *taking pictures of students' conceptual visualizations* using workshop materials such as pin boards, presentation cards, flip charts, stickers and posters;

3) *oral reflections* of students on given tasks and exercises within the tutorial at several points of time;

4) *written hindsight* of students after the tutorial reflecting upon the exercises carried out, the methods used and specifically the emotions they felt during learning processes.

Interactions were mainly analyzed by several discussion rounds among the facilitators of the tutorials. In addition, the students' work was analyzed with qualitative content analysis. While doing content analysis is more or less uncritical, since it happens after the course, reflecting on observed interaction in this context can become quite challenging since the role of the course facilitators is extremely volatile. Jumping between the two roles as participating instructor and as observing researcher may engender cognitive erasures and biases in observations and results caused by capacity limitations of human working memories (cognitive load). So, when taking over the instructor's role not every incident may be observed and remembered and when taking over the observer's role not every hint could instantly be given on the shop floor that might be useful for enabling improvements.

8.5. Results

In general, it can be stated that students used the tasks, preparing a pitch with creativity techniques and presenting them to an audience, to test themselves in different roles. During the feedback stage of the pitches, all students had to slip into the different roles of entrepreneurs presenting and defending their own pitch and of investors to comment on the business idea and give suggestions for improving the pitches of others.

8.5.1. *Developing, presenting and defending a pitch*

In the tutorials for the faculty of business and economics, there was no group consisting exclusively of engineering students, so that a best possible stimulus for multiperspective real-world problem solving by means of engineering methods could be ensured for engineering students. Within the presented concepts of engineering creativity in section 8.2.2, the students from the faculty of business and economics mainly developed engineering ideas as technical solutions for solving real-world problems. In their solutions the students proposed technical artifacts and media services, which have the potential for incremental but also for disruptive improvement of the related fields. According to the 6F model, they were especially encouraged to go for F6 "imagining new ideas". The 10 most interesting cases are briefly presented in the following:

1) "Hover shoes" – smart shoes to individualizing the public traffic system;

2) "WETAIR" – a diving mask to filter the oxygen directly from surrounding water;

3) "Sunny Win$" – transparent window layer for energy generation and for isolating of temperature fluctuation;

4) "beer lover" – producing biological and durable beer as unique drinking entertainment providing novel flavors;

5) "Flexy" – instantly drying textile fabrics (e.g. for avoiding changing wet swimwear on the beach);

6) "iBot" – a massage robot for head, neck and back;

7) "Your car, your mat" – a foot mat cover for cars that can be changed quickly and easily to bring cleanliness and fresh air into the car;

8) "Auxilio – The social sharing platform" – a free of charge app for sharing social and service;

9) "Digital Lifestyle Consulter" – an app to serve for facilitating on decision making;

10) "Linky" – an app that spontaneously links people together for parties and events.

Even if there were groups without engineering students, there were still a fair number of engineering ideas on their tables during the first phase of the idea generation. However, the further selection process of these groups showed that the non-engineers did not have the confidence to try engineering ideas, even if they could have subcontracted the necessary engineering competences.

Concerning the quality of the developed concepts, almost all students' ideas were based on idealism. Nearly all students wanted to make the world a better place. Sustainability and social responsibility were very important issues for them. However, they had great problems to propose their business ideas in a way that they would be profitable. While presenting the pitch, entrepreneurs with idealistic ideas and an underestimation of the importance of financing had to face sharks serving the only economic objective to make as much money as possible. This discord of contradicting anticipations led to lively pitches and insightful learning experiences.

In contrast to that, the project management students from the engineering education department of the faculty of mechanical engineering got the tangible task to prototype explicit extensions for the learn-to-walk toy car, especially by addressing F4 "making" on the bases of F6 "imagining new ideas". The eight interesting ideas are as follows:

– Mechanical extensions

1) "WaWe3000" – water gun with water tank and pump, blue light and stickers in the police design (see Figure 8.2);

2) "Easy Stack" – forklift attachment;

3) "MogKu" – modular muted clutch with sweep system;

4) "MB sKids" – skids with footboard and support straps for an accompanying person.

– Electrical and electronic extensions:

1) “The Kit” – light and sound, rewritable board as a number plate;

2) “DaProTech” – impact protection with warning lamp and warning sound;

3) “Lumos” – lighting kit, personalized design;

4) “MB Car Sense” – transmission of steering movements to control a video game, distance measurement to objects/persons, melodies and lighting by push-button printing (see Figure 8.3).

It is particularly interesting that at the level of concrete product development, half of the ideas are not mechanical engineering ideas, although they were developed by students of mechanical engineering, but focused on ideas from the Education for Innovation and Technology (EIT) sector.

Figure 8.2. *“WaWe3000” – water gun with water tank and pump, blue light and stickers in the police design*

In both cases, the more open and visionary scenario in the business and economics tutorial, as well as the more tangible prototyping scenario in the faculty of mechanical engineering, the application of the Shark Tank Experience led to appropriate results regarding the learning objectives and ILOs. Concerning the roleplay, students liked to experience roles and related perspectives of both, being entrepreneurs and being investors. Students stated that it was very difficult to bear criticism while being in the entrepreneurs' role. Since they somehow have identified themselves with their ideas in the meantime, they took investors' critique on their ideas personally. Nevertheless, they also said at the end of the tutorial that they would never again underrate the significance and meaning of finances in this context. Students who started to play the "sharks" said they had an additional benefit, as they already knew what is vital from the investors' viewpoint. They were able to fine-tune their pitch based on playing the investor's role, when it was their turn to play the entrepreneur's role.

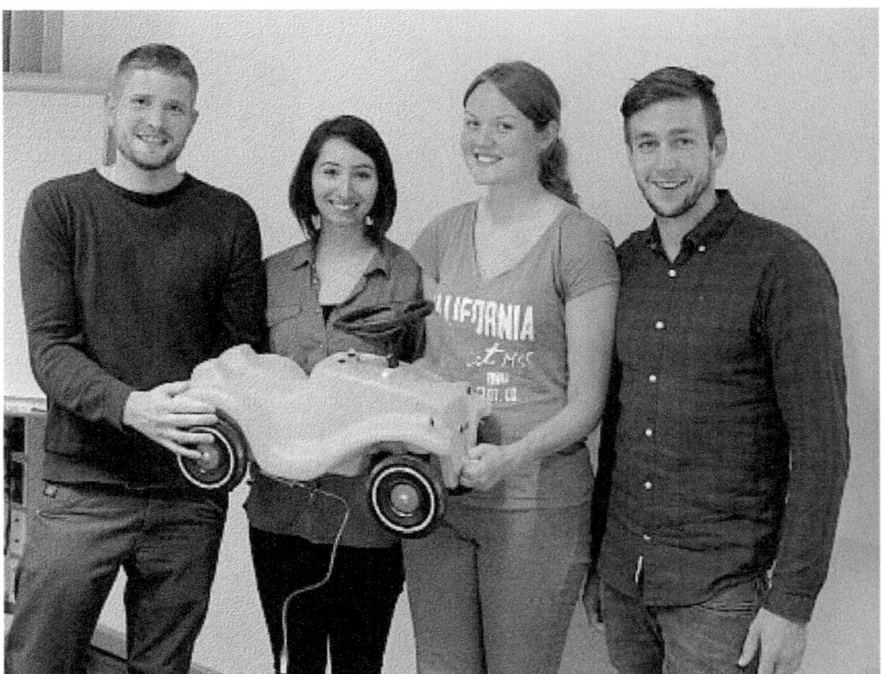

Figure 8.3. *"MB Car Sense" – transmission of steering movements to control a video game, distance measurement to objects/persons, melodies and lighting by push-button printing (realization with Arduino)*

8.5.2. *Do something unusual!*

Due to the structure of the tutorial as a 2-day workshop, the students had to fulfill the "do something unusual" task without comprehensive planning between the end of the workshop on day 1 and its beginning on day 2. In general, it can be stated that all students faced this task mainly with curiosity. However, they also had severe concerns that their ideas to solve the task may be too unspectacular for the other students. After the course facilitators relentlessly repeated that no idea is too unspectacular and no one should ask too much of himself or herself, the task was accepted by the majority. Furthermore, the task should not be solved with a simple test of courage like bungee jumping but should give to the students the possibility to deal with each ones' individual willingness to be creative. Hence, the solutions varied a lot due to students' diverse backgrounds and personalities. For example:

– one student started a conversation with an unknown cashier in a supermarket;

– one student decided to take off his shoes and walked the long way home without wearing them;

– one student dialed a phone number she did not know to overcome her fears of talking to strangers;

– one student went shopping in his pajamas;

– one student decided to walk backwards for a while in the university's canteen;

– one student played songs in the pedestrian mall and collected money for that;

– yet another student asked to take away for free the already prepared meal when it came to pay in a diner.

More or less, all students found a sufficient solution that fitted their individual context despite their fears of the uncertain and inconvenient situation. Reflecting the experiences in the following discussion, most students said that they had found a personal profit by solving this task. When asked to describe their feelings they had perceived before, during and after solving their tasks, a typical process with individual differences was observable. During the development of the idea and the decision to make it happen, a phase of increasing unease became stronger and stronger reaching its peak right before the actual implementation of the idea. During its actual implementation, the students had a feeling of unease that changed into focusing on the actual situation, own actions and the reactions of the surrounding. After accomplishing the task, the students noticed a time of relaxation and a feeling of pride of having overcome the challenge.

8.5.3. *Formative evaluation*

As a reaction on the students' restrained risk taking even in a benevolent pedagogical setting, several changes have been made from one tutorial to the other: the theoretical part on creativity and innovation has been reduced to the most necessary. The reflection resulted in inserting a second cycle in the syllabus of the tutorial: while the first round is now more to get to know the full cycle and therefore ideas for incremental improvements have been appreciated as well as ideas for disruptive improvement, the second one is reserved for really visionary and disruptive ideas. For this reason, in the second round the facilitators constantly tried to motivate the students by friendly encouragement convincing them to be free to choose whatever they want. During idea generation by applying divergent thinking they should even ignore natural laws, consider an unlimited budget and all other conceivable constraints if it would help create a really grand and disruptive idea. They should simply keep any objections for the convergent thinking phase. As a result, the ideas were more and more in line with expectations, especially during the last run in the summer term 2017.

8.6. Discussion

Above all, the Shark Tank Experience concept used in the tutorial has proved to be very effective for the students in terms of experiencing entrepreneurial thinking and creativity. It provides learning occasions in terms of gaining relevant knowledge, understanding and developing the personality of students based on practical experiences and reflection of them.

– All students could experience that the application of creativity techniques for idea generation led to reasonable results. Moreover, while additionally integrating the Business Model Canvas led to better business cases due to negotiating its inherent systemic constraints, the more open version led to stronger visionary pitches concerning disruptive potentials.

For integration of both principles attributing them the same importance more time than 2 days would be necessary.

– The roleplay situation simulated the most vital point in creative and entrepreneurship procedures: the pitch. Pitching determines life or death of ideas and it is an imperative step from creativity to innovation. In this phase, the uncertainty is as high as possible. It is navigation in open systems at its best.

– The small group size in combination with an active workshop atmosphere was very appropriate for the context. The small numbers of participating students allowed for in-depth discussions and made it possible to address and foster each student individually, similar to a coaching process.

– The facilitators checked out several group sizes between two and five members. However, there is no identifiable ideal group size for fostering creative teamwork in this setting, yet. Beyond that, the need to agree on a shared business idea fostered profound teamwork.

– All six facets of the Dortmund 6F-model to foster creativity in higher (engineering) education – reflective learning (F1), autonomy: independent self-determined learning (F2), curiosity and other intrinsic motivation (F3), making: learning by creating (F4), multiperspectives (F5) and imagining new ideas (F6) – could be addressed as learning objectives with the presented Shark Tank Experience scenario.

In summary, all participating students asserted a high level of satisfaction and intrinsic motivation at least for parts of the tutorial. This can be considered a success in contrast to many other course scenarios in higher engineering education based at best on extrinsic motivation. However, students (engineering as well as economics and business students) surprised the tutorial facilitators by avoiding radical innovations and risk taking even in a playful atmosphere.

Overall, without emphatic and ongoing encouragement, students avoided radical solutions in both tasks of the shark tank experience (the pitch and doing something unusual), no matter what subject they studied.

– To this end, a great amount of time of the tutorials had to be spent on working with students' willingness and confidence to take risks, even in these playful situations.

– The real difficulty for the teachers is to challenge the students' risk tolerance in their generation of ideas and even more in their choice of ideas without taking the decision out of their hands. The dividing line between convincing and persuading is narrow.

– Regarding the "doing something unusual" task, some students managed it and some struggled and got by anyway, but only after reflecting on their fears and making sure that more simple solutions are valued as well.

– Dealing with uncertainty was a challenging experience for most of the students, as they stated in the feedback phases, even though the course facilitators told repeatedly that not the quality or quantity of their ideas was graded.

Why is that? If universities in particular and maybe societies and their educational systems in general do not foster and support non-conformity, risk taking and endurance of uncertainty, but still stick to the conformity pressure and social control of the hidden curriculum, students may not show it on demand and remain initially suspicious – even in playful simulations. Moreover, if learning from failure and from Aha-moments is something that is desired by every teacher in theory, but finally undesired in practice and, moreover, graded poorly because it does not coincide with the official knowledge and its inherent conformity pressure, students simply unlearn daring to be creative and innovative.

As an example, Haertel *et al.* [HAE 12] show that the engineering curricula of German technical universities do not intend to foster independent learning and working, multiperspective thinking or the development of original ideas.

As another example, Beghetto's studies [BEG 10] of students' creativity avoiding practices in the classroom reveal teacher-related and organizational aspects like:

– prevalence of teacher-dominated convergent teaching approaches; teachers' need for sticking to the plan and placing value on the acquisition of facts;

– teachers' view that unexpected student ideas are disruptive; even soon-to-be teachers generally prefer expected ideas over unexpected or unique ideas;

– wrong assumptions about the role of creativity in the classroom and wrong expectations about students' motivation;

– scripted curricula that represent the most extreme form of convergent teaching, separating learning from the development of creative thinking;

– teaching for the test and increased use of externally mandated, fact-based tests.

In particular, teaching for the test (in combination with convergent teaching and facts acquisition orientation) highlights to students what is really valued and important [BIG 11, GUI 50]. Needless to say that teachers' prior experiences and personal success stories during their own school and university years with those approaches can lead to the wittingly or unwittingly reproduced practices across time, place and persons [BEG 10]. Moreover, Schank and Childers identify that a too critical attitude in widespread academic competition, rivalry and paternalism seem to hamper the unfolding of the creative attitude: "Indeed, in our society showing why someone's idea won't work is one of the hallmarks of erudition" [SCH 88]. Instead, supporting students' creativity and innovativeness would demand teachers to be creative [HAE 15]. As a result, the inclusion of creativity, entrepreneurship and innovativeness in higher engineering education curricula demands a transformation of mind sets and personality on the part of teaching staff to contribute to adjustments in the engineering syllabus [BYE 13].

Moreover, there seems to be another paradox: How is it possible to gain those personality effecting long-term oriented learning outcomes in a temporary context of a common short-term university course consisting of fifteen 2 h lessons within three or four months? There is no easy solution for this problem, and a lot of future work has still to be done. A first step as a bottom-up approach is to ask students to repeat the task "Do something unusual" by themselves regularly, and using each run of it to go further small step by small step. Finally, on an organizational level, a concept for "innovation mainstreaming" [HAE 17b] to support creativity, entrepreneurial thinking and innovativeness has to be developed and implemented in the educational system.

Finally yet importantly, May *et al.* ask whether entrepreneurship is an innate ability, or a skill that can be acquired through education and is rather influenced by socialization. From their point of view, society significantly influences the willingness to become entrepreneurial. Cultures that value entrepreneurship have more entrepreneurs than societies in which entrepreneurship is not fostered or even valued. This indicates that the environment influences entrepreneurial behavior at least to a certain degree [MAY 15].

8.7. Prospective work

Based on the successful effects of the Shark Tank Experience in both the entrepreneurship tutorial and the project management course, the concept of fostering students' creativity in disciplinary contexts will be expanded in the winter term 2017–2018. To come next, besides the two already existing learning scenarios, the tutorial to the seminar "imparting of technology" in the mechanical engineering faculty will be enriched by creativity training and an adaption of the Shark Tank Experience. Since the students in this course are going to be school teachers in technical subjects, the presentations will be somehow different to the pitches in the entrepreneurship tutorial and project management course. In this case, the soon to be teachers will be asked to develop innovative learning scenarios that foster their students' curiosity and enthusiasm for technology. These scenarios will be tested and evaluated in real-life situations in schools or university school labs. Furthermore, the engineering labs in which the students work during the project management course and the imparting of technology course will provide materials and instructions that allow students to easily apply the creativity techniques they already learnt in the tutorials.

8.8. Conclusion

Modern societies are not only in need of creative, innovative and entrepreneurial engineers, they are also in need of engineers having the courage to take risks and realize their ideas. The Shark Tank Experience tutorial presented a learning scenario on how it is possible to learn to be more creative and innovative, or in short thinking like an entrepreneur. Providing engineering students with a setting allowing for acting in this way should be an overall task of engineering education on every organizational level from the single lesson to the course, to the curriculum to the department to the whole university and to the entire educational system for aligning engineering education on innovation mainstreaming. However, there are important character traits that students cannot learn by simply applying a straightforward learning path under the common time regime of common higher engineering education, such as courage, risk taking, non-conformity, openness and going against the mainstream. These observations refer to the gap between knowing how to make something become an academic reality (generate ideas, design a business model in a tutorial) and how to make something become a reality in the real world (transforming these ideas into real businesses).

It is important to have these insights in mind, if all the contemporary and prominent statements echoing around the globe about the need for engineers, who represent the profession that solves the world's most pressing problems preferably from the technical perspective, are serious suggestions and not just symbolic bustle.

Finally, students should not wait for their institutions or their teachers to give them permission to be creative and innovative. Instead, they should simply start alone. To this end, they should somehow practice each day:

1) to reflect on their personal creative sets and design their settings accordingly;

2) to use creativity techniques to generate ideas;

3) to show steadily courage to express themselves and to achieve their ideas.

Probably, this is all it takes.

8.9. Acknowledgments

The presented work was done in the scope of the national research projects "ELLI – Excellence in Teaching and Learning in Higher Engineering Education" (part 1: 2011–2016, part 2: 2016–2020) funded by the German Federal Ministry of Education and Research. Pioneering work was done in the scope of "Da Vinci – Fostering Creativity in Higher Education" (2008–2011) funded by the German Federal Ministry of Education and Research and "TeachING-LearnING.EU – Competence Centre for Engineering Sciences" (2010-2014) funded by Mercator Foundation and Volkswagen Foundation.

Additional work has been done in the scope of the international research project “Innov’ing 2020 – les ingénieurs et l’innovation: nouveaux métiers, nouvelles formations” (2015–2017) funded by L’Agence nationale de la recherche de la France–ANR.

Very special thanks to Silke Frye, Benedikt Schwuchow, Siegmar, Oehland, Julian Mosler and André Funke from the Engineering Education Department at TU Dortmund University. In addition, we also would like to thank Sebastian Gajewski from the Institute for Design and Materials Testing at TU Dortmund University for, spontaneously and out of competition, developing a prototype of a doner kebab vertical rotisserie extension enhancing the learn-to-walk toy car.

Last but not least, we thank all students from Dortmund as well as to all Erasmus students from Italy, Poland and Portugal, who participated very actively in the courses and contributed their creative ideas as well as their courage to do something unusual.

8.10. References

[ASM 08] ASME, available at: www.asme.org/, 2008.

[AUL 13] AULET B., URSACHE M., *Disciplined Entrepreneurship: 24 Steps to a Successful Startup*, John Wiley & Sons, Hoboken, 2013.

[BAB 15] BABBIE E.R., *The Practice of Social Research*, Cengage Learning US, Boston, 2015.

[BEG 10] BEGHETTO R.A., “Creativity in the classroom”, in KAUFMAN J.C., STERNBERG R. J. (eds), *The Cambridge Handbook of Creativity*, Cambridge University Press, Cambridge, 2010.

[BEL 16] BELSKI I., ADUNKA R., MAYER O., “Educating a creative engineer: learning from engineering professionals”, *Procedia CIRP*, vol. 39, pp. 79–84, 2016.

[BER 07] BERTRAM U., PREIßING W. (eds), *Navigieren im Offenen System: Kunst – Transfer – Management*, Container-Verl., Filderstadt, 2007.

[BIG 11] BIGGS J. B., TANG C., *Teaching for Quality Learning at University: What the Student Does*, McGraw-Hill Society for Research into Higher Education & Open University Press, Maidenhead, 2011.

[BRO 11] BRODIN E. M., FRICK L., “Conceptualizing and encouraging critical creativity in doctoral education”, *International Journal for Researcher Development*, vol. 2, no. 2, pp. 133–151, 2011.

[BUL 95] BULL K. S., MONTGOMERY D., BALOCHE L., "Teaching creativity at the college level: a synthesis of curricular components perceived as important by instructors", *Creativity Research Journal*, vol. 8, no. 1, pp. 83–89, 1995.

[BYE 13] BYERS T., SEELIG T., SHEPPARD S. *et al.*, "Entrepreneurship: Its Role in Engineering Education", *The Bridge*, vol. 43, no. 2, pp. 35–40, 2013.

[CDI 17] CDIO, available at: www.cdio.org/, 2017.

[CHR 00] CHRISTENSEN C.M., OVERDORF M., "Meeting the challenge of disruptive change", *Harvard Business Review*, March–April 2000.

[CLA 00] CLARKE A.C., *Profiles of the Future: An Inquiry into the Limits of the Possible*, Indigo, London, 2000.

[CRA 14] CRAWLEY E.F., MALMQVIST J., ÖSTLUND S. *et al.*, *Rethinking Engineering Education: The CDIO Approach*, Springer, New York, 2014.

[CRO 01] CROPLEY A., *Creativity in Education and Learning: A Guide for Teachers and Educators*, Routledge, London, 2001.

[CRO 05] CROPLEY D.H., CROPLEY A.J., "Engineering creativity: a systems concept of functional creativity", in KAUFMAN J. C., BAER J. (eds), *Faces of the Muse: How People Think, Work and Act Creatively in Diverse Domains*, Lawrence Erlbaum, Hillsdale, 2005.

[CRO 10] CROPLEY D.H., CROPLEY A.J., "Functional creativity: "products" and the generation of effective novelty", in KAUFMAN J.C., STERNBERG R.J. (eds), *The Cambridge Handbook of Creativity*, Cambridge University Press, Cambridge, 2010.

[CRO 11] CROPLEY D.H., KAUFMAN J.C., CROPLEY A.J., "Measuring creativity for innovation management", *Journal of Technology Management and Innovation*, vol. 6, no. 3, pp. 13–30, 2011.

[CRO 16] CROPLEY D.H., "Creativity in engineering", in CORAZZA G. E., AGNOLI S. (eds), *Multidisciplinary Contributions to the Science of Creative Thinking*, Springer, Singapore, 2016.

[DEH 12] DEHNART A., "Shark Tank's sneak attack!", *The Daily Beast,* November 5, 2012, available at www.thedailybeast.com/shark-tanks-sneak-attack.

[DOW 05] DOWNEY G., "Are engineers losing control of technology?", *Chemical Engineering Research and Design*, vol. 83, no. 6, pp. 583–595, 2005.

[DRU 02] DRUCKER P., "The discipline of innovation", *Harvard Business Review*, August 2002.

[EDI 13] EDISON H., ALI N. B., TORKAR R., "Towards innovation measurement in the software industry", *Journal of Systems and Software*, vol. 86, no. 5, pp. 1390–1407, 2013.

[FIN 10] FINK A., GRABNER R. H., GEBAUER D. *et al.*, "Enhancing creativity by means of cognitive stimulation: evidence from an fMRI study", *NeuroImage*, vol. 52, no. 4, pp. 1687–1695, 2010.

[GAR 96] GARFINKEL H., *Studies in Ethnomethodology*, Polity Press, Cambridge, 1996.

[GOF 10] GOFFMAN E., *Behavior in Public Places: Notes on the Social Organization of Gatherings*, Free Press, New York, 2010.

[GOL 06] GOLDBERG D. E., *The Entrepreneurial Engineer: Personal, Interpersonal, and Organizational Skills for Engineers in a World of Opportunity*, John Wiley & Sons, Hoboken, 2006.

[GOL 92] GOLEMAN D., KAUFMAN P., RAY M. L., *The Creative Spirit*, Dutton, New York, 1992.

[GUI 50] GUILFORD J. P., "Creativity", *American Psychologist*, vol. 5, no. 9, pp. 444–454, 1950.

[HAE 12] HAERTEL T., TERKOWSKY C., JAHNKE I., "Where have all the inventors gone? Is there a lack of spirit of research in engineering education curricula?", *15th International Conference on Interactive Collaborative Learning (ICL)*, Villach, Austria, September 26–28, 2012.

[HAE 15] HAERTEL T., TERKOWSKY C., RADTKE M., "Creative students need creative teachers: Fostering the creativity of university teachers: a blind spot in higher engineering education?", *2015 International Conference on Interactive Collaborative Learning (ICL),* Florence, Italy, September 20–24, 2015.

[HAE 16a] HAERTEL T., TERKOWSKY C., "Creativity versus adaption to norms and rules: a paradox in higher engineering education", *International Journal of Creativity & Problem Solving (IJCPS)*, vol. 26 no. 2, pp. 105–119, 2016.

[HAE 16b] HAERTEL T., TERKOWSKY C., MAY D., "The Shark Tank Experience: How Engineering Students Learn to Become Entrepreneurs", *2016 ASEE Annual Conference & Exposition,* New Orleans, USA, 26-29 June 2016.

[HAE 16c] HAERTEL T., TERKOWSKY C., OSSENBERG P., "Kreativtität in der Hochschullehre: ‚Tue etwas Ungewöhnliches!'", in HEINER M., BAUMERT B., DANY S. *et al.* (eds), *Was ist "Gute Lehre?" Perspektiven der Hochschuldidaktik*, Bertelsmann, Gütersloh, pp. 73–92, 2016.

[HAE 17a] HAERTEL T., TERKOWSKY C., JAHNKE I., "To develop creative students, you may have to redesign yourself as well as your teaching", *Creative Academic Magazine*, February 2017.

[HAE 17b] HAERTEL T., TERKOWSKY C., WILKESMANN U., "Innovation Mainstreaming in der Hochschule: Wie kommt das Neue in die Lehre?", in BÜCKER D., DANDER V., GUMPERT A. *et al.* (eds) *"Trendy, hip und cool": auf dem Weg zu einer innovativen Hochschule?*, Bertelsmann, Gütersloh, pp. 55–62, 2017.

[HOF 12] HOFFMANN K., *Dein Mutmacher bist du selbst!: Faustregeln zur Selbstführung*, Springer, Dordrecht, 2012.

[JAC 16] JACKSON N., "Discussion paper", *Creative Academic Magazine,* no. 7, October 2016–2017, available at: www.creativeacademic.uk/uploads/1/3/5/4/13542890/discussion_paper.pdf.

[JAH 17] JAHNKE I., HAERTEL T., WILDT J., "Teachers' conceptions of student creativity in higher education", *Innovations in Education and Teaching International*, vol. 54, no 1, pp. 87–95, 2017.

[JØR 11] JØRGENSEN U., BRODERSEN S., LINDEGAARD H. *et al.*, "Foundations for a new type of design-engineers", *ICED 11 The 18th International Conference on Engineering Design*, Copenhagen, Denmark, August 15–19, 2011.

[LEM 17] LEMAÎTRE D., "Current developments in the French engineering education system", *European Journal of Engineering Education*, vol. 42, no. 2, pp. 131–141, 2017.

[LO 11] LO C. T., The impact of entrepreneurship education on entrepreneurial intention of engineering students, PhD Thesis, City University, Hong Kong, 2011.

[MAY 15] MAY D., HOSCH-DAYICAN B., LEISYTE L. *et al.,* "Entrepreneurship and gender in higher engineering education in Germany", *43rd Annual SEFI Conference*, Orléans, France, June 29–July 2, 2015.

[MAY 94] MAY R., *The Courage to Create*, Norton, New York, 1994.

[MOR 17] MORACE C., MAY D., TERKOWSKY C. *et al.*, "Effects of globalisation on higher engineering education in Germany: current and future demands", *European Journal of Engineering Education*, vol. 42, no. 2, pp. 142–155, 2017.

[NAT 04] NATIONAL ACADEMY OF ENGINEERING, *The Engineer of 2020: Visions of Engineering in the New Century*, National Academies Press, Washington DC, 2004.

[NIU 09] NIU W., LIU D., "Enhancing creativity: a comparison between effects of an indicative instruction "to be creative" and a more elaborate heuristic instruction on Chinese student creativity", *Psychology of Aesthetics, Creativity, and the Arts*, vol. 3, no. 2, pp. 93–98, 2009.

[OST 10] OSTERWALDER A., PIGNEUR Y., *Business Model Generation: A Handbook for Visionaries, Game Changers, and Challengers*, John Wiley & Sons, Hoboken, 2010.

[PLU 04] PLUCKER J. A., BEGHETTO R. A., DOW G. T., "Why isn't creativity more important to educational psychologists?: Potentials, pitfalls, and future directions in creativity research", *Educational Psychologist*, vol. 39, no. 2, pp. 83–96, 2004.

[SCH 88] SCHANK R. C., CHILDERS P. G., *The Creative Attitude: Learning to Ask and Answer the Right Questions*, Macmillan, New York, 1988.

[TER 13] TERKOWSKY C., HAERTEL T., BIELSKI E. *et al.*, "Creativity@School: mobile learning environments involving remote labs and E-portfolios. A conceptual framework to foster the inquiring mind in secondary STEM education", in DZIABENKO O., ZUBIA J.G. (eds), *IT Innovative Practices in Secondary Schools: Remote Experiments*, pp. 255–280, University of Deusto, Bilbao, 2013.

[TER 16] TERKOWSKY C., HAERTEL T., ORTELT T. *et al.*, "Creating a place to bore or a place to explore? Detecting possibilities to foster students' creativity in the manufacturing engineering lab", *International Journal of Creativity & Problem Solving (IJCPS)*, vol. 26, no. 2, pp. 23–45, 2016.

[TSA 13] TSAI K.C., "A review of the effectiveness of creative training on adult learners", *Journal of Social Science Studies*, vol. 1, no. 1, pp. 17, 2013.

[WIL 12] WILSON W. B., *Redefining Success: Still Making Mistakes*, Penguin, Toronto, 2012.

9

Engaging with Heritage to Promote Innovative Thinking in Engineering Management Education

For those colleagues responsible for educating future engineers and engineering managers, the question of how to promote innovative thinking among engineering students within a rigidly structured curriculum, which is often driven by professional body requirements and built around intended learning outcomes and performance indicators, is difficult to reconcile. In focusing on a new pedagogical approach to teaching engineers, this chapter argues that by "thinking out of the box" and by setting learning outside of the "traditional" classroom and boundaries, engineering educators can both promote innovative thinking and enhance the student experience. This chapter discusses a graduate level management module taught as part of a number of master's programs in engineering and engineering management. It provides a unique insight into engineering education in the UK and concludes by suggesting that while innovative thinking cannot be directly taught to students, in leading by example, engineering lecturers can act as role models; nurturing and encouraging students step out of their routine "comfort zones" and find new ways of working in which innovation and creative thinking becomes an everyday reality.

Chapter written by Jane ANDREWS and Robin CLARK.

9.1. Introduction

Defining innovative thinking in engineering as the "the ability to address challenges and opportunities by the creative application of engineering capital, knowledge and theory", this chapter is set within a postmodern society whereby professional engineers are expected to be able to provide innovative and bespoke solutions to an array of social, environmental and scientifically created problems. Within this setting, the question of how to provide an education which balances the multidimensional requirements of society with sound engineering theories and principles demands a paradigm shift within engineering education and across the engineering sector [NĂD 16, BRO 14]. Indeed, at a time when engineers are expected to be multiskilled, yet highly specialized "*Professional Practitioners*" able to manage a multidisciplinary team of people with a range of capabilities and knowledge, the question of how to provide an education that promotes innovative thinking skills and leadership abilities within individual students is one which continues to allude many of those tasked with educating future engineers.

One of the consequences of this is that, finding themselves unprepared and ill-equipped to deal with management issues, engineering graduates often seek to "upskill" themselves by enrolling on "traditional" graduate business management programs. Such programs, which include those leading to a master's in business administration (MBA) qualification, tend to be purposefully constructed so as to meet the industrial training needs of a multitude of managers representing the complete spectrum of occupational classifications. As such, they equip students with generic management skills based on general management theory. From an engineering perspective, one of the key weaknesses with this approach is that in appealing to a cross-section of managers, the pluralistic nature of postmodern society is often ignored, particularly when it comes to the engineering sector and the needs and demands placed upon today's engineers. Indeed, while business school graduate education may be suitable for those working in the business sector, it is not unreasonable to suggest that it is unsuitable for professional engineers who find themselves in a management role. Put another way, business school higher education lacks engineering context and application and is thus not ideal for engineers [FRI 03]. Conversely, an opposing view is that the generic nature of business school graduate education is exactly what engineering managers need, as it provides students with transferable skills relevant to *any* management setting [GRE 04]. This second argument is not held by the chapter authors, who believe that engineers need to be trained to manage and lead engineering projects in an engineering school where their specialist knowledge and skills are best understood and nurtured and where the engineering context can be created.

Taking a step back from this argument, it is reasonable to assume that the majority of graduate engineers will, at some time in their professional careers, find themselves in a management role. Furthermore, the need to make sure that professional engineers are adequately prepared to step into a management role means that engineering education has a responsibility to provide suitable training that is both appropriate to, and contextualized within, an engineering setting. While some engineering schools do attempt to do this, the problem of how to provide a solid engineering-focused management training program that encourages student engineers to apply innovative thinking skills to management problems remains largely unaddressed. Hence, the question of "What is the best pedagogical approach to adopt when training engineers to become managers?" is, up to this point in time, generally unanswered.

In seeking to address the above question, this chapter builds on previous work to consider graduate engineering students' perceptions and experiences of a project management module that is embedded into a number of master's programs within a school of engineering. The module develops and promotes leadership skills and competencies, including high levels of communication and innovative thinking skills, within an engineering setting. It was constructed around an approach to engineering education that was developed by the authors with the specific intention of promoting scholarship in engineering education [CLA 14]. Termed the RVS Model of Engineering Education, the approach is based upon three distinctive yet interrelated themes (relationships, variety and synergy). This chapter introduces the concept of synergetic configuration. Part of a much wider empirical approach to engineering education, the work discussed in this chapter draws upon the findings of an exploratory study. It builds upon previous publications and knowledge to discuss how the relationships, variety and synergy (RVS) approach to engineering education may be used to promote innovative thinking in students.

9.2. Background: the importance of engineering education

In a report published some 18 months before the Brexit vote, the UK Secretary of State for Business, Innovation and Skills, Vince Cable (MP), argued that "A strong British engineering sector is vital to the long term sustainability of our economic recovery, and increasing the supply of engineers is at the heart of this" [CAB 15]. Since this time, much has changed on the socioeconomic and political front. In June 2016, a referendum on membership of the EU unexpectedly resulted in a small majority of the UK public voting to leave the EU. While the ramifications of this unprecedented decision have yet to unfold, there is little doubt that as an isolate *sitting* on the edge of the European Union, the UK will need innovative thinkers more than it ever has before. Indeed, at a time when "engineering skills shortages" are frequently reported in the UK media, there is some suggestion that a lack of engineering talent potentially threatens future national and international economic success and security [ONI 15, GRO 14, PER 13, WAL 17]. It is important to stress that the need for innovation in engineering is not restricted to the UK or even the EU, irrespective of country or continent; higher education needs to produce innovative thinking professional engineers able to play a full role in solving transnational problems and challenges across all areas of contemporary life [NAE 17]. From providing solutions to global problems (which vary in nature from pollution to terrorism for example) through to satisfying what can only be described as an insatiable global public demand for innovative technology, the fact is that engineers are central to our future and, as such, sit at the heart of global society.

Set against this challenging environment, engineering educators need to produce "well-rounded" professional graduate engineers, who are able to apply innovative and appropriate mathematical and scientific theory to engineering problems, while managing what is more often than not a cross-disciplinary team of people [SHA 02, KUM 07, TOO 11, CAR 16]. The reason behind is that this is not complicated. Engineering projects are best led by professional engineers. While this may seem like common sense; it is not unreasonable to postulate that across the globe in any one day there are thousands, if not millions, of engineering projects that are ultimately managed by individuals who do not hold a qualification in engineering and who have little or no understanding of underpinning epistemologies. Such individuals, many of whom are professionally trained managers, can only base engineering decisions on management indicators, including financial outputs and time pressures. Unfortunately, as tragic events in London in June 2017 highlighted the human cost and socioeconomic impact of allowing people who are not professional engineers to make engineering decisions can be catastrophic [BOO 17, DON 17].

In seeking to equip engineering students with management skills, many engineering schools have introduced business modules into engineering programs. This chapter focuses on one such module, *Project Management for Engineers*. This graduate-level course uses the concept of engineering heritage to equip future engineering managers with transferable project management skills including innovative thinking. Offered as part of a master's in engineering or engineering management, the module is built upon an innovative approach to engineering education developed by the authors and as such incorporates three key concepts, relationships, variety and synergy [CLA 14]. This chapter considers in some depth an evaluation of the module. It introduces the concept of synergetic configuration as a way to promote innovative, management practices including innovative thinking among engineering graduates.

9.3. Synergetic configuration: an innovative approach to engineering education

Building on previous work published by the authors, the concept of synergetic configuration emerged out of the RVS approach to engineering education [CLA 14]. Defined as "the requirement that university level Engineering Education should equip students with the technical capabilities required by industry, while providing them with a sound theoretical knowledge base, within a supportive curriculum in which 'softer skills' are embedded alongside technical competencies and independent learning is the norm", synergetic configuration involves a synthesis and application of three distinctive concepts, relationships, variety and synergy. Originally articulated to make the concept of scholarship in engineering education more accessible to engineering educators, the RVS approach is constructed in such a way so as to ameliorate the many linguistic and conceptual barriers engineering colleagues encounter when attempting to engage with pedagogic theory.

Grounded in work published by Boyer [BOY 90], the RVS approach encapsulates four distinctive but interconnected areas of scholarship (discovery, integration, application and teaching) [BOY 90]. In synthesizing each of these four areas with the three constituent parts of the RVS approach, the authors of this chapter have developed a pedagogic approach that not only produces innovative thinkers, but it also enhances the engineering student experience.

Closely aligned with innovative thinking in engineering, the first component of scholarship that of discovery is represented in engineering education by the pursuance of new knowledge acquired through teaching and research. This component aligns closely with the concept of variety within the RVS approach as it encompasses new ways of thinking and doing "engineering education".

From here, the multidisciplinary nature of engineering means that the next component of scholarship, integration is particularly relevant to engineering education; it is encapsulated within the RVS model through the concept of synergy. Professional engineers have to draw upon and synergistically apply a range of different theories and approaches in their daily work. A central component of the RVS approach, the scholarship of integration promotes synergy in engineering education by encouraging students and educators to look across and beyond traditional disciplinary boundaries, applying innovative thinking to problems in a way that is *imaginative*, *interdisciplinary*, *interpretive* and *integrative* [BOY 90, p. 21].

The third area of scholarship application combines with the relational aspect of the RVS approach to become the cornerstone of engineering education. By its very nature, engineering is an applied discipline. Engineers need to be able to identify, interpret and integrate the various relational linkages, which exist across the theory–practice/socioengineering nexus. In doing so, they need to be able to build relationships across and between diverse groups of people, many of whom have little or no understanding of engineering theory or practice. The ability to develop and maintain strong working relationships, irrespective of external context or pressures, and based upon trust and respect is vital for engineering managers. It is not enough to be able to apply theory; engineering managers need to be able to communicate such theory in a manner that is fully accessible and in doing so provide a level of assurance that they are skilled professionals who can be trusted. It is the responsibility of engineering educators to make sure that they are equipped to do this.

The final area of scholarship, that of teaching, is central to all three of the RVS concepts and so represents the fundamental purpose of the RVS model, which is to provide an accessible and adaptable, scholarly approach to educating future engineers. Teaching engineering needs not only to be based on innovative thinking; it also needs to promote innovative thinking among the students. By bringing together the three distinctive concepts of relationships, variety and synergy, this chapter builds on the RVS approach to consider how the proactive synergetic configuration of pedagogy, practice and procedure ($\boldsymbol{P}^3$) within different elements of a graduate engineering management program can act to promote innovative thinking. In doing so, it considers how heritage may be used to promote a positive learning experience at the graduate level, equipping future engineers with vital management skills and competencies.

9.3.1. *Heritage, innovation and project management: the learning and teaching context*

In applying the RVS approach to graduate level engineering management education, the project management module uses heritage engineering to contextualize contemporary engineering practice and procedure while encouraging students to consider how innovative thinking in engineering impacts the world in which we live. Changing the heritage site from year to year, the teaching materials are purposefully constructed so as to link the past with the present, and engineering with society. Over the past 5 years, the following heritage sites have been used:

1) *SS Great Britain*: Located in Bristol, the SS Great Britain is a Steam Ship, built by Brunel and launched in 1843. Pedagogically, the site provides an ideal example of innovative thinking in engineering. The learning materials associated with this site were constructed so as to enable engineering students to gain an understanding of the history of project management in engineering while considering the contemporary engineering and management challenges associated with recovering, restoring and preserving the ship (for more information, see http://www.ssgreatbritain.org/story).

2) *Manchester Museum of Science and Industry (MOSI)*: Situated in the city of Manchester, MOSI houses a range of historic artifacts, buildings and innovations dating back some 250 years. The learning materials associated with this site contextualized innovation in engineering within an engineering project management setting, encouraging students to investigate the breadth and depth of the challenges engineering managers have faced over the past three centuries (for more information, see http://msimanchester.org.uk/collection).

3) *RAF Cosford*: Based in the county of Shropshire, RAF Cosford is the home of the RAF Museum. It comprises a large range of military artifacts dating back around a century. Housed in purpose-built hangers and specialist buildings, from a pedagogical perspective this site provides the ideal context to teach students about the role of engineering innovation in maintaining peace and also, on the negative side, in developing weapons of mass destruction. It also provides the means by which contemporary engineering and management challenges around health and safety, preservation and restoration can be explored. This was captured, in particular, by exploring the Dornier restoration project ongoing at the museum (for more information, see https://www.rafmuseum.org.uk/cosford/).

4) *The Brunel Museum*: Located in London, the Brunel Museum is the site of the first ever tunnel that was built underneath the River Thames. Pedagogically, it provides an excellent example of engineering innovation, enabling students to consider how "blue-skies" thinking by an engineer in the 19th Century continues to impact contemporary engineering theory and practice across the globe, and in doing so impacts all of our day-to-day activity. This example has many elements that blend the technical and managerial challenges that are faced in making a project a reality (for more information, see http://brunelmuseum.dyndns.info/).

5) *Derwent Valley Mills*: Situated in the Amber Valley in the county of Derbyshire, Derwent Valley Mills is a UNESCO World Heritage Site. It is from this site that the "factory system" of manufacturing was first developed and operated. This site has much to offer from a pedagogical perspective, providing the means by which students are able to consider the ethical issues around engineering innovation while encapsulating management challenges around manufacturing, civil and mechanical engineering. The juxtaposition of old and new in terms of the development of the site is made very apparent to the students at this site (for more information, see http://www.derwentvalleymills.org/).

Each of the above heritage sites offers a unique setting in which students were encouraged to explore the importance of applying innovative engineering thinking skills to project management. The module assessment was synergetically configured so as to encapsulate and apply relevant engineering and management theories and practices, while providing "real" examples of how innovative thinking in engineering can change the world. In seeking to continually improve and innovate, the module is evaluated annually, with the findings from each annual evaluation being fed into the development of the next year's teaching materials. The following sections are based on an exploratory study centered upon an evaluation of the most recent site to be visited, Derwent Valley Mills.

9.3.2. *The P^3 Project: methodology*

Naming the project after the three main elements of academic practice identified above, pedagogy, practice and procedure (P^3) and adopting an Action Research Approach [NOR 09], the P^3 Project set out to investigate the following research problem:

> "How can pedagogy, practice and procedure be improved through the application of the RVS model of Engineering Education at graduate level in a Module which uses Heritage Engineering to contextualize Engineering Management issues?"

A survey tool was administered to a cohort of 80 engineering students from a range of engineering disciplines, all of whom were enrolled upon the engineering project management module. The response rate was 61 students, 75% of the sample.

Incorporating the concepts that make up the RVS approach, the survey was divided into three main areas: relationships, variety and synergy. Each of these concepts aligns with one of the three areas upon which the P^3 study was focused: practice (relationships), pedagogy (variety), procedure (synergy).

9.3.2.1. *The P^3 Project sample*

In putting together the learning and teaching materials for the module, it was important to take account of the fact that the students hail from a number of different engineering disciplines, ranging from chemical and civil engineering through to mechanical and electronic engineering including many others. Some of the students were enrolled in a master's in engineering program (MEng), whereas others were enrolled in a master's of science in engineering management (MSc Eng Mgmt). Figure 9.1 shows that under two-thirds of the cohort was enrolled in the MEng program, with the rest, all qualified engineers in possession of at least a bachelor's level qualification, studying for an MSc in an engineering management related discipline.

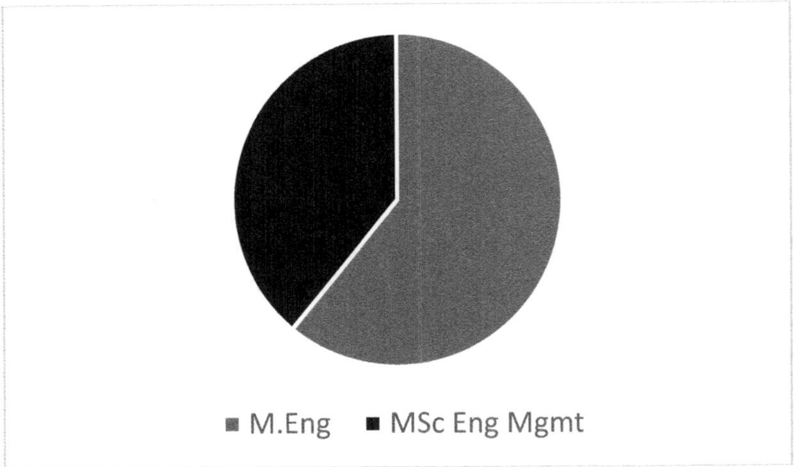

Figure 9.1. *Program of study: percentage of sample (N = 61)*

In looking closely at the demographic makeup of the sample, it was noted that under a quarter of the sample were female; this is typical of the school of engineering in which the study was conducted, but atypical of wider UK society whereby between 6 and 9% of the engineering workforce are women.

Gender	*N*	%
Male	47	77
Female	14	23
Total	61	100

Figure 9.2. *Demographic breakdown of sample: gender*

In considering the issues around engineers who are employed as managers, one of the key variables that emerged out of the literature is related to individual work experiences, with those engineers who had undertaken a period of work experience during their training being more successful in their careers. Figure 9.3 shows that the majority of the sample had undertaken a paid internship as part of their undergraduate training. This figure, which for the UK is quite high, is indicative of the School of Engineering & Applied Science where the study was carried out, whereby all undergraduates are strongly encouraged to participate in a year-long formal work experience.

Work experience	%
Undergraduate internship	54
Part-time work while studying	9
Engineering management role	8
Engineer	5
Consultant engineer	2
Other	7
None	15
Total	100

Figure 9.3. *Sample work experience*

9.3.3. *Innovating pedagogy and practice: the P^3 Project study findings*

In seeking to provide an overview of the manner in which engineering heritage may be used to promote innovative thinking in engineering, the P^3 Project study findings are now discussed. It should be noted that the decision was taken at the beginning of the study not to attempt to directly measure students' perceptions of how their ability to "think innovatively" was impacted by the project management module. There are a number of reasons for this, the most important one relating to the psychological epistemology underpinning the concepts of "thinking" and "innovative thinking". While both of the researchers are experts in engineering education, neither has any expertise in educational psychology. Thus, innovative thinking which was defined as "the ability to address challenges and opportunities by the creative application of engineering capital, knowledge and theory" is implicitly examined in this chapter through a number of different lens, each relating to the RVS approach and contextualized within an engineering education setting.

9.3.3.1. *Innovation and relationships in training engineering managers*

Perhaps the most important aspect of the RVS approach is that it places importance on developing a practice-based approach to learning and teaching in engineering. This is particularly important given that working in teams and developing relationships with work colleagues are generally accepted to be prerequisites to innovative thinking and innovation in practice [SCH 10, BEC 07]. This is particularly important for engineering students, who, upon graduating, often quickly find themselves managing teams of people from a wide range of disciplinary backgrounds bringing with them different levels of knowledge and understanding.

The student-focused pedagogy that is central to the RVS approach necessitates the building of positive work–study relationships among groups of students as well as between students and lecturers. Working in teams represents an important part of learning, particularly when almost inevitably tensions and problems arise. While students look to their lecturers to act as role models and show them how to behave in difficult situations, it is important that they "learn by doing" and experience first-hand how to manage disparate teams of people who often have conflicting perspectives. Such skills are invaluable in the workplace.

In seeking to evaluate the "relationship" aspect of the RVS approach, seven relational practice-focused question–statements were asked. While a four-point Likert scale was used in all of the questions, the differentiation between the levels of agreement and disagreement was not valid and hence the "agree and strongly agree" data (denoted as A) and "disagree and strongly disagree" data (denoted as DA) are merged together enabling a clearer picture of the students' perspectives.

Figure 9.4 below provides an overview of the percentage of those students who either agreed or disagreed with the statement–questions.

In thinking about the project management module the following applies...	A	DA
The group work in this module has enabled me to build some close friendships	47	11
The group work has provided the opportunity for me to develop my communication skills	53	4
The Heritage fieldtrip has provided the opportunity for me to get to know my classmates better	48	10
I have found the module lecturer approachable throughout	54	2
I found it difficult to work in a group	20	39
The opportunity to gain feedback in stages has helped me develop my understanding of PM	49	6
The class visit by a project manager helped me appreciate the real-world PM context	55	1

Figure 9.4. *Relational (practice-focused) questions*

In looking at the above findings, the most surprising result relates to the fact that only 20 of the sample indicated that they found it difficult to work in a group. While this is a relatively high percentage of the sample (33%), it is lower than would be expected given the widely discussed difficulties with group work reported both anecdotally within the university and identified in the literature [TUC 16, MOO 15, WEA 14].

One of the key aims of the project management module is to use the relationship concept within the RVS model to establish a learning environment in which innovative thinking was promoted. In considering the relational nature of the RVS approach, students' positive disposition both toward working in groups and participating on a fieldtrip to a heritage site (on this occasion Derwent Mills) suggests that the module provides the ideal environment for innovative thinking to be nurtured.

An important part of the pedagogical approach is continual feedback and feedforward, this is offered so as to encourage students to develop as they progress through the module. Forty-nine of the samples found this helpful, which is another indictment of the value of the approach in nurturing innovative thinking.

9.3.3.2. *Innovation and variety in pedagogic practice*

Having examined the issues around innovation and relationships in teaching and learning practice, the survey then focused on the variety element of the RVS approach. Exposure to a variety of different experiences, scenarios and situations not only engender innovative thinking among students, it also provides an interesting and relevant learning experience [MIS 13, MAR 06, CAL 04].

One of the key aspects of the project management module upon which the evaluation is focused is the manner in which the management of engineering projects has changed over time. This is contextualized by the fieldtrip to the heritage site, in this case Derwent Mills, whereupon students were encouraged to look at the different socioeconomic, engineering and political issues associated with engineering project management from the end of the 18th Century through to modern-day practice. Questions asked in this section of the survey were, therefore, focused on classroom activities but also included a look at students' perceptions of the value of the fieldtrip. Figure 9.5 outlines the seven question statements that were asked. The data are again disaggregated into "Agree" (A) and "Disagree" (DA).

In thinking about the project management module the following applies...	A	DA
The different learning and teaching approaches used in the module made the lectures more interesting	57	0
The different learning and teaching approaches used in the module made the content more understandable	56	1
The use of case-studies has helped me appreciate the range of practical project management issues	55	2
The class discussions have been valuable in helping me understand the main issues	50	6
The lectures provided the foundational knowledge needed for me to study the subject independently	57	4
The visit to Cromford Mill was interesting in helping me understand the concept of industrial heritage	50	6
The presentation at Cromford Mill was useful in helping me comprehend some of the practical issues associated with PM	60	1

Figure 9.5. *Variety in pedagogy questions*

While working in teams represents an important pedagogical tool in nurturing innovative thinking, a central part of the RVS approach is to promote independent learning. The variety of pedagogies that students were exposed to in the module not only included a fieldtrip, but also involved class discussions, other active learning exercises as well as case study learning. The above survey findings suggest that the use of a wide range of learning approaches was viewed positively by the students. Indeed, having provided foundational knowledge within a positive learning environment, the module assessment then called for a level of applied theoretical knowledge combined with innovative and independent thought. That all but three students (out of a cohort of 90) passed the module gives a clear indication that this approach was successful.

9.3.3.3. *Innovating synergy in learning and teaching*

The final area explored by the survey related to the manner in which the module leader had purposefully synergistically constructed the learning materials to encapsulate all aspects of the students' current and future ontology, epistemology and future employability. The purpose of doing this was to provide a learning environment in which innovative thinking could be both role modeled by the lecturer and nurtured in the students. Figure 9.6 provides an overview of the disaggregated data in the form of 10 questions related to synergy.

In thinking about the project management module the following applies...	A	DA
The module content is relevant to modern day engineering	60	1
The module content is applicable to the coursework	60	1
The group work in the module is good preparation for work	57	4
The real life case studies helped me understand the theory	53	8
The visit to Cromford Mill helped me gain a wider perspective on project management	46	9
The visit to Cromford Mill has given me an insight into the importance of Heritage within contemporary society	51	7
The visit to Cromford Mill brought project management to life	51	10
The module learning outcomes have been achieved	60	1
The assessment is appropriate for developing my project management skills	58	3
The module content is relevant for a career in industry	59	2

Figure 9.6. *Synergy in pedagogic practice questions*

The findings of the survey provided positive evidence that the module lecture's attempts to provide a synergetic experience were valued by the students. The use of an action research approach means that the areas which were not identified as positively as the others (relating to the contextualization of contemporary project management in terms of the fieldtrip) will be addressed next year, whereupon more attention will be given to linking heritage and history with contemporary management and engineering.

9.3.3.4. *RVS as a route to innovative thinking in engineering*

The interpretation and application of the evaluation discussed in this chapter provides an interesting overview of how, through the use of an innovative approach to learning and teaching, students can be encouraged to develop into innovative thinking engineering managers. While this chapter only looks at one module, it should be noted that the RVS approach forms the strategic underpinning for the learning and teaching strategy in the engineering school in which this work was based. In developing an accessible and fully transferable pedagogical approach specifically for an engineering education context, the authors have seen eight different cohorts graduate and go onto to excel in industry.

9.4. Moving forward: the application of RVS to achieve synergetic configuration and student success

The use of an action research approach combined with the application of the RVS approach has provided an excellent learning environment in which the intended learning outcomes were achieved through synergetic configuration of the curriculum. In seeking to provide an accessible representation of this, a conceptual framework has developed a diagrammatic representation, which is shown in Figure 9.7.

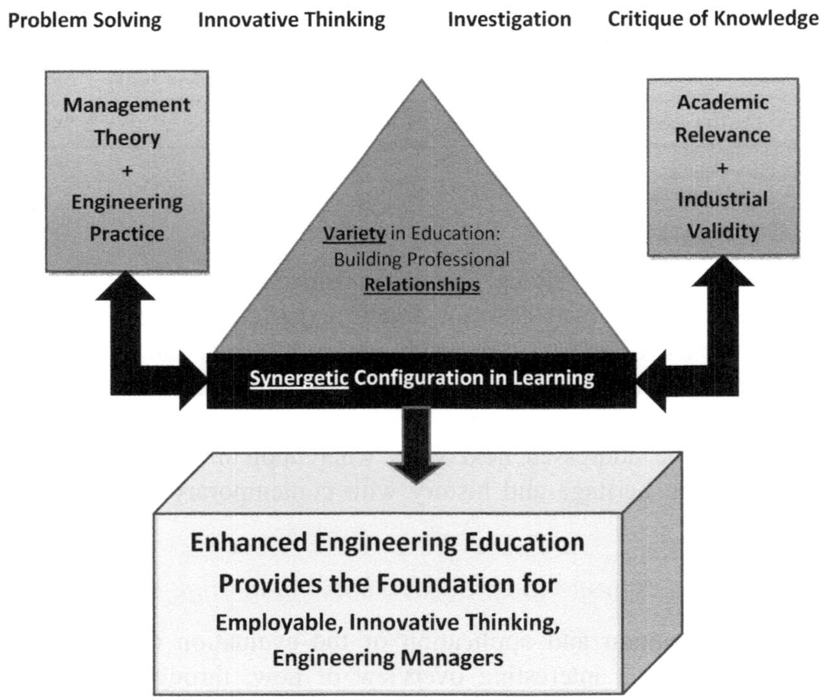

Figure 9.7. *Emergent conceptual framework: synergetic configuration in engineering management education*

Building upon the evaluation findings, the conceptual framework depicted in Figure 9.7 shows the centrality of the main three components of the learning approach proposed by Clark and Andrews [CLA 14]: *relationships, variety and synergy*. In developing this approach, the authors applied the findings of a number of educational research studies that they had jointly led over a number of years; the largest one of which examined the experiences of around 1,000 undergraduate students and found that the most important driver of student success is a "sense of belonging" [AND 12, CLA 12, AND 13, CLA 13]. It is this sense of belonging that is at the heart of innovation in learning and teaching and central to the nurturing of innovative thinking in students. When students feel they belong at an institution, they are more likely to work at a higher level. For engineering students, working at a higher level naturally involves innovative thinking. Indeed, the application of the RVS approach combined with synergetic configuration of the curriculum today can do much to nurture the engineering managers of tomorrow. Yet this is only the beginning. Developing innovative thinking in engineering students is not about applying the RVS approach to the curriculum in a discrete way; it also involves embedding the synergetic configuration throughout the curriculum – in the materials the students engage with, the people they interact with and in the work they are guided to perform. The use of heritage in learning is one way of doing this.

Increasingly, when we look at engineering as a discipline or profession, the increasingly diverse expectations that employers have of new engineering graduates are simply not possible in one individual – even after successful completion of an intensive 3-, 4- or 5-year program of study. Yet there is little doubt that within contemporary society, engineering has truly become "a Team Profession", with different individuals bringing different skills, experiences, perceptions, ideas and knowledge to create the whole. This need for synergetic working leads to the idea of an "engineering professional" rather than someone who is simply an "engineer" [CLA 17] working in isolation.

The concept of the engineering professional suggests that innovative thinking skills are now more important than ever. Indeed, conceptualizing "engineers" as "engineering professionals" has the potential to really open "eyes, hearts and minds" and in doing so help engineering to tackle some of the "internal challenges" it currently faced by across the globe (including issues around gender, difficulties attracting new engineers, concerns regarding the influence of professional bodies on the curriculum, the need for more effective industrial and employer involvement in the engineering education curriculum to name but a few). Furthermore, the development of high-level innovative thinking skills also has much to offer with regard to the development of individuals capable of making an effective contribution to future society.

9.5. Conclusion

This work has presented heritage as the vehicle for change but there are clearly many other possibilities. What is evident is that through dialogue, awareness and co-creation in its many forms, engineering and engineering management can foster innovative thinking and create an environment in which engineering professionals can truly impact our world for the better.

In conclusion, for those colleagues responsible for educating future engineers and engineering managers, the question of how to promote innovative thinking among engineering students within a rigidly structured curriculum is one that is difficult to address. By focusing on the RVS approach to engineering education and through introducing the concept of synergetic configuration, this chapter has shown how, by “thinking out of the box” and outside of the “traditional” laboratory or classroom, engineering educators can both promote innovative thinking and enhance the student experience. While the example given in this chapter relates to the use of heritage engineering as a learning and teaching tool, examples of innovative thinking in engineering are evident in every aspect of our day-to-day lives. By acting as role models and by adopting innovative teaching methods which stretch and challenge students, engineering lecturers can nurture and encourage future engineers, encouraging individual students to step out of their routine “comfort zones” and apply engineering capital to everyday challenges. In this way, engineers will continue to lead innovation in our society, inventing new technologies and providing solutions to as yet unknown problems in a world that only they, as engineers, can begin to imagine.

9.6. References

[AND 12] ANDREWS J., CLARK R., THOMAS L., *Compendium of Effective Practice in Higher Education Retention & Success*, Higher Education Academy, York, 2012, available at: http://www.heacademy.ac.uk/assets/documents/what-works-student-retention/What_Works_Compendium_Effective_Practice.pdf.

[AND 13] ANDREWS J., "Editorial: breaking down the barriers", *Journal of Widening Participation and Lifelong Learning*, vol. 14, pp. 1–3, 2013.

[BEC 07] BECKMAN S.L., BARRY M., "Innovation as a learning process: Embedding design thinking", *California Management Review*, vol. 50, no. 1, pp. 25–56, 2007.

[BOO 17] BOOTH R., "Tower cladding tests after Grenfell fire lack transparency, say experts", *The Guardian*, 2017, available at: https://www.theguardian.com/uk-news/2017/jun/26/grenfell-tower-cladding-linked-to-fire-pulled-from-sale-worldwide.

[BOY 90] BOYER E.L., *Scholarship Reconsidered: Priorities of the Professoriate*, Princeton University Press, Lawrenceville, NJ, 1990.

[BRO 14] BROOKS C.K. (ed.), *Beyond Postmodernism: Onto the Postcontemporary*, Cambridge, Cambridge Scholars Publishing, 2014.

[CAB 15] CABLE V., "Foreword", *The State of Engineering in the UK*, Engineering UK, London, 2015.

[CAL 04] CALLAN V., *Building Innovative Vocational Education and Training Organisations*, National Centre for Vocational Education Research Ltd., Adelaide, Australia, 2004, available at: http://files.eric.ed.gov/fulltext/ED493934.pdf.

[CAR 16] CARTER D.F., RO H.K., ALCOTT B. *et al.*, "Co-Curricular Connections: The Role of Undergraduate Research Experiences in Promoting Engineering Students' Communication, Teamwork, and Leadership Skills", *Research in Higher Education*, vol. 57, no. 3, pp. 363–393, 2016.

[CLA 12] CLARK R., ANDREWS J., "Peer mentoring in higher education: a reciprocal route to student success", in ANDREWS J., CLARK R., THOMAS L. (eds), *Compendium of Effective Practice in Higher Education Retention & Success*, Higher Education Academy, York, pp. 72–75, 2012.

[CLA 13] CLARK R., ANDREWS J., THOMAS L. *et al.*, *Compendium of Effective Practice in Higher Education Retention & Success: Vol. II*, Higher Education Academy, York, 2013.

[CLA 14] CLARK R., ANDREWS J., Relationships, variety & synergy [RVS]: the vital ingredients for scholarship in engineering education? A case-study, *European Journal of Engineering Education*, vol. 39, no. 6, pp. 585–600.

[CLA 17] CLARK R., ANDREWS' J., *Engineering or the Engineer? A Paradox of Professionalism*, IET Symposium on Innovation in Higher Education, London, 2017.

[DON 17] DONNOLLY L., "NHS alert as fire chiefs warn 38 hospitals are 'as dangerous as Grenfell Tower'", *The Independent,* 2017, available at: http://www.telegraph.co.uk/news/2017/06/27/nhs-alert-fire-chiefs-warn-38-hospitals-dangerous-grenfell-tower/.

[FRI 03] FRIGA P.N., BETTIS R.A., SULLIVAN R.S., "Changes in graduate management education and new business school strategies for the 21st century", *Academy of Management Learning & Education*, vol. 2, no. 3, pp. 233–249, 2003.

[GRE 04] GREY C., "Reinventing business schools: the contribution of critical management education", *Academy of Management Learning & Education,* vol. 3, no. 2, pp. 178–186, 2004.

[GRO 14] GROOM B., "Shortages of engineers threaten UK growth warn employers", *Financial Times*, 2014, available at: http://www.ft.com/cms/s/0/3d3e3218-174c-11e4-b0d7-00144feabdc0.html#axzz3RLIvhJhx.

[KUM 07] KUMAR S., HSIAO J.K., "Engineers learn 'soft skills the hard way': planting a seed of leadership in engineering classes", *Leadership and Management in Engineering*, vol. 1, pp. 18–23, 2007.

[MAR 06] MARR W.A., "Geotechnical engineering and judgment in the information age", *GeoCongress 2006: Geotechnical Engineering in the Information Technology Age*, pp. 1–17, 2006.

[MIS 13] MISHRA P., FAHNOE C., HENRIKSEN D., "Creativity, self-directed learning and the architecture of technology rich environments", *TechTrends*, vol. 57, no. 1, p. 10, 2013.

[MOO 15] MOORE P., "Cultural diversity in group work", 2015, available at: http://www.uow.edu.au/dvca/ltc/teachdev/octal/content/groups/public/@web/@cedir/documents/doc/uow098312.pdf.

[NĂD 16] NĂDRAG L., BUZARNA-TIHENEA A., "From Modern to Postmodern curriculum", *The Annals of UOC: the Philology Series*, 2016, available at: http://litere.univ-ovidius.ro/Anale/documente-z/art-rez-2015-2/10_Nadrag-Galbeaza%202-2015.pdf.

[NAE 17] NAE, *Grand Challenges for Engineering*, 2017, National Academy for Engineering, Washington, available at: *http://www.engineeringchallenges.org/*.

[NOR 09] NORTON L.S., *Action Research in Teaching and Learning: A Practical Guide to Conducting Pedagogical Research in Universities*, Routledge, London, 2009.

[ONI 15] ONITA L., "Shortage of Engineering Skills to Cost UK £27bn a Year", *Engineering & Technology Magazine,* 2015, available at: http://eandt.theiet.org/news/2015/jan/skills-shortage-engineering.cfm.

[PER 13] PERKINS J., *Review of Engineering Skills*, Department of Business, Innovation & Skills, London, 2013, available at: https://www.gov.uk/government/uploads/system/uploads/attachment_data/file/254885/bis-13-1269-professor-john-perkins-review-of-engineering-skills.pdf.

[SCH 10] SCHUHMANN R.J., "Engineering leadership education–the search for definition and a curricular approach", *Journal of STEM Education: Innovations and Research*, vol. 11, nos. 3–4, pp. 61, 2010.

[SHA 02] SHAW W.H., "Engineering management in our modern age", *2002 IEEE International Findings of the Engineering Management Conference, 2002 (IEMC'02)*, pp. 504–509, 2002.

[TUC 16] TUCKER R., "Prejudicial evaluation: bias in self-and-peer-assessments", *Collaboration and Student Engagement in Design Education*, vol. 12, p. 76, 2016.

[TOO 11] TOOR S.U.R., "Differentiating leadership from management: an empirical investigation of leaders and managers", *Leadership and Management in Engineering*, vol. 11, no. 4, pp. 310–320, 2011.

[WAL 17] WALLACE T., TOVEY A., "UK firms from engineers to banks struggle to fill skills shortage", *The Telegraph*, 2017, available at: http://www.telegraph.co.uk/business/2017/03/04/uk-firms-engineers-banks-struggle-fill-skills-shortage/.

[WEA 14] WEAVER C.L., DURAN E.C., NIKLES J.A., "An integrated approach for development of scientific writing skills in undergraduate organic lab", *Addressing the Millennial Student in Undergraduate Chemistry*, pp. 105–123, 2014.

10

How Do Graduate Engineering Schools Train for Innovation? Study of the Curricula of Three French Schools

10.1. Introduction

As the first chapters of this book show, conceptions of innovation vary according to socioeconomic contexts, professional activities and epistemological approaches of the disciplines that account for the phenomenon. What about the training itself, when it actually comes to training engineers? It seems that different conceptions of innovation also affect educational practices. The incentive to innovate, which has been present in our societies for a few decades, in fact receives a mixed response and is interpreted differently by higher education institutions training engineers. It reinterprets representations of the role of engineers in relation to progress and its place and role in society since the Renaissance, Enlightenment and the Industrial Revolution (see Chapter 1).

How do engineering schools adapt the incentive to innovate and how do they combine it with their educational principles? What does it mean, from a practical point of view, to want to train for innovation? What are the methods used and the preferred curricular orientations? What are the links between educational innovation and innovation training for engineers?

Chapter written by Denis LEMAÎTRE and Christophe MORACE.

This chapter attempts to answer these questions in a practical way based on the survey carried out within the Innov'ing 2020 research project: *Innov'ing 2020: Les ingénieurs et l'innovation: nouveaux métiers, nouvelles formations*. It consists of an analysis of the literature devoted to innovation in the curricula of French schools, then the case study of three schools that are particularly advanced in this field. The purpose of this survey is to highlight the breadth of pedagogical practices and identify current trends in innovation training. The French case is quite specific, because of the history of schools and the notion of the type of engineers to which they refer. However, beyond this specificity, the problems mentioned and the curricular orientations observable are of a rather general scope. The chapter begins by presenting the French context and the conceptions of innovation, before detailing three case studies of French schools specialized in innovation, then proposing, in a wider context, a modeling of the logic behind these training programs.

10.2. The adaptation of French engineering schools to innovation

10.2.1. *The sociohistorical context*

The theme of innovation is relatively recent, given the history of engineering training in France. The implementation of training was made, starting from the 17th Century, following the creation of King's engineer corps, whose goal was to bring the technical advances necessary for economic development, border protection and the kingdom's expansion. This idea of engineers at the service of the State thus inspired the creation of the first schools (École nationale supérieure des mines, École des ponts et chaussées), and then the École polytechnique in 1794. In the 19th Century and with the appearance of the Industrial Revolution, arose the need to train engineers to serve the industry and private companies. A number of schools were created in France in order to meet the needs of locally established nascent industries, as a result of discoveries and technical progress (coal and steel, textiles, chemistry, etc.). The French Graduate engineering schools, unlike those of other countries, are thus created outside the universities, according to this dual model consisting of serving both the interests of the State and nascent industries. It was not until the 20th Century, after the Second World War, that university-type schools were more widely created in France, with the aim of democratizing access to the profession and increasing the number of graduate engineers. This explains the large number of engineering courses (200 accredited for awarding degrees at Master's 2 level), most of them outside the classic university system, and some not covered by the French Ministry of Higher Education, but other ministries (Ministries of Agriculture, Industry, Armies), which contributes to the complexity of the French system. In terms of curricula, a double orientation dominates: high-level scientific training (mathematics, physics, biology) and a strong historical link with companies [LEM 17]. Thus, in its own way, the French model responds to the desire to create higher education for the training of elites capable of national development and furthering progress, a function that Whitehead [WHI 29] attributed to higher education, summarizing the ideas of his time.

The multiplicity of schools and their links with industries gives rise to a relatively large diversity of structures, often of small size (a few hundred students for the most part), and consequently benefiting from a great deal of pedagogical freedom. This small institution size and their relative autonomy allows field teams to design curricula adapted to the ambitions of schools and industrial fields, by being inventive in terms of teaching methods, particularly in relation to university courses designed for students in large numbers. From this point of view, French schools enjoy a fairly high capacity for adaptation, stimulated today by the needs of industrial innovation and new contemporary technical challenges.

Schools must take into account changes in the French industrial landscape and changes in the careers and jobs of their graduates. Until the end of the 20th Century, graduate engineers from schools were destined to work for, for the most part, large industrial groups (automotive, aeronautics, pharmaceutical, agribusiness, energy, etc.), in close relation to national history. Since then, a strong trend toward internationalization has changed the structure, location and recruitment of these industrial groups. At the same time, deindustrialization and the rise of digital technologies are encouraging the development of small businesses and new ways of conceiving the employment of engineers, many of whom are now working in service companies. This change in context has prompted schools to rapidly develop training in innovation and entrepreneurship, and to place more importance on the creation of industrial activities, notably through research and the creation of incubators. This is largely embodied by the term innovation in the context of French engineering schools.

10.2.2. *The conceptions of innovation beginning with curricular changes*

The incentive to innovate is relayed by the *Commission des titres d'ingénieur* (CTI), the sole agency that accredits engineering schools and gives them the right to award degrees in France on the basis of specific standards and official recommendations. Let us recall that in France there is only one category of graduate engineers, corresponding to a Master's 2 level. This agency plays a central role in the engineering curriculum in France. In its recommendations, the CTI clearly indicates its expectations as concerns innovation, linking it to the idea of entrepreneurship (see Chapter 6), which presents the results of a survey on the evolution of schools in the field. In the "reference documents", published for schools[1] by the agency, the topics on which these schools are evaluated are explained, and therefore constitute guidelines to follow in the design of training programs. Among the "major criteria for accreditation", innovation is mentioned in the school's "missions", in a subsection entitled "innovation, development, transfer and entrepreneurship". This title illustrates the fact that innovation is considered here as the development catalyst of new tools, objects or processes, intended to be transferred to companies and the world of industrial production, which confirms this recommendation: "The school contributes by its educational and research activities in the creation of innovative projects, products or services, activities and enterprises". In the criteria relating to training, innovation is based on abilities and intellectual posture:

> The opening up of the future engineer towards innovation and the creation of activities or enterprises is based on the development of a state of mind, the realization of activities and a training process. These approaches should encourage divergent thinking, the right to make mistakes, risk taking, the exercise of creativity and critical thinking, consideration of the user's needs and customer's requirements, the pedagogical capitalization of experiences, trans-disciplinary openness, and commitment to entrepreneurship, curiosity and openness to the world [SIC 17].

1 https://www.cti-commission.fr/fonds-documentaire/document/15/chapitre/242?a=1, accessed July 26, 2017.

In this case, we have a fairly fundamental and broad approach to innovation such as intellectual capacity, which is centered on creativity and critical thinking, and a field of application, which is centered on the market and customer relationships. On the pedagogical level, the training objectives are translated in the form of recommendations for specific activities:

> The opening is concretized by transversal activities and specific events allowing any student to conduct a personal or collective creation project (innovation or activity) allying notably heeding needs, creativity, experimentation, and the conception of a business plan [SIC 17].

These criteria encourage pedagogical methods that are far removed from traditional disciplinary methods (courses, tutorials or practical work) and are more oriented toward trans-disciplinary approaches and active pedagogy, which confront students with needs relating to the professional world. The mention of a "business plan" links innovation to the creation of industrial activities in market logic, through entrepreneurship.

The CTI's requirements reflect the broader societal expectations of engineering education and their economic, social and industrial roles. In this entrepreneurial approach to innovation, we can see the reflection of utilitarian design that today dominates French engineering schools. The links between industry and schools lead to the conception of innovation as being inseparable from technical applications and the resolution of circumscribed problems [FOR 11]. Societal, cultural, social and human forms of innovation still have little relation to technical innovation in schools. In the CTI's documents [SIC 17], however, a chapter is devoted to societal aspects, under the title "Sustainable development, social responsibility, ethics and deontology", but it is separate from that on innovation and there is no explicit link between the two. This conception of innovation, in relation to market logic and to industrial development, largely dominates in French engineering schools.

This is confirmed by a study conducted by the French Engineering Education Study Committee in 2013, "Innovation Training in Engineering Schools"[2]. From the beginning of the document, which proposes to explain the concept by means of a literature review and practices, the author

2 http://www.cefi.org/cefiblog/espace-membresecole/dossiers/, accessed July 26, 2017.

indicates in the introduction p. 11 that “from the set of definitions we can already retain three axioms: (1) no innovation without sanction by the market, (2) no innovation without design (object, product, services), (3) no innovation without an innovative enterprise”. The document emphasizes the link between the incentive to innovate and the need for industrial recovery and a revitalization of the economy through business creation.

Should we keep in mind that any form of innovation encouraged in engineering courses is systematically associated with the placing of new products or services on the market? An in-depth review of the schools' websites, visits to schools and reading the available literature shows that the situation is much more varied in the field. Within schools, many activities dedicated to innovation are developed independently of this logic. In schools where research is directed toward the production of more fundamental knowledge (for example in physics, applied mathematics, computer science), pedagogical activities can encourage students to produce innovative concepts without thinking of marketable industrial applications. Some approaches to innovation also lead to thinking about design that go beyond the immediate logic of the market, such as sustainable development, education and disability. Finally, some schools encourage, for example through student extracurricular activities, humanitarian type projects, in which the innovations proposed by the students, often to the benefit of disadvantaged countries that they visit, aim to help populations in difficulty.

Graduate engineering schools thus evolve according to their environment and reference frameworks that can be contradictory: research and production of knowledge, industrial activity and the market, sociopolitical norms, the social role given to engineers and training ethics. The ways of conceiving innovation, in terms of educational objectives, reflect these contradictions. It is therefore retained that, despite this conception of innovation as the production of new goods or services in market logic, there are other approaches within schools that we need to explore further through pedagogical practices.

10.2.3. *Forms of pedagogical innovation*

Examining the context of schools shows us that the pedagogy of innovation is strongly linked to the pedagogical innovation as such. To train for innovation, teachers and educational leaders are convinced of the need for a pedagogical breakthrough, as a trigger, a change in classic academic practices (knowledge-centered lectures) and a form of teaching that provides authorization and incitement to intellectual and moral postures allowing novelty.

The survey carried out during the Innov'Ing 2020 project on innovation training in French engineering schools included a study of the available literature on pedagogical practices, particularly through the proceedings of the symposia on *questions relating to pedagogy in higher education*, school websites, publications and journal articles published on the subject. A number of principles are emerging that characterize the pedagogy of innovation. They can be summarized according to three general themes.

The first theme that crosses educational thought is that of disciplinary decompartmentalization. The curricula of French engineering schools are strongly structured by academic disciplines. These different disciplines are a reflection of the skills that we want to transmit to graduate engineers. First, they concern the technical specialty in which the students are trained, whether it is, for example, agronomy, public works or computer science. Within a specialty corresponding to an engineering profession, different technical disciplines intervene. For example, the training of an engineer specialized in shipbuilding can include, besides general mechanics, structure mechanics, fluid mechanics, material mechanics and so on. The general development of technical knowledge leads to a diversification of the disciplinary fields within the training programs, which is inevitably found in the curricula, when it comes to training experts in mastering their field. In addition to these specialty disciplines, there are other disciplines in the professional development of engineers, such as management, law, modern languages and the humanities. Disciplinary diversification and partitioning are further reinforced by the academic movement of schools [LEM 07], following the recruitment of research lecturers and the development of research since the 1990s. In the end, engineering curricula appear as sorts of disciplinary patchworks. This explains why, in French schools, the educational effort is largely concerned with disciplinary groupings and the development of cross-disciplinary activities. Innovation, as shown in Part 1 of this book, is often the result of shifting concepts from one field to another, or the overcoming of contradictory imperatives from different aspects of a problem (economic, technical and environmental). Training for innovation thus supposes decompartmentalizing disciplinary fields through transversal mechanisms, outside of classic academic teaching. These systems can be located in schools, but also in cooperation between different higher education institutions. We thus obtain methods for gathering together civil engineering students and architecture students, or engineering students and business school students, sometimes even on an international scale.

Another theme running through innovation training is that of openness to the socioprofessional reference practices pertaining to engineers in companies. As we saw above, the CTI strongly insists on this idea of openness in its recommendations to schools. Historically speaking, French engineering schools have been established as a response to the activity of specific industrial sectors, identifiable with specific regions and industries such as aeronautics in Toulouse or chemistry in eastern France. Many schools have set up a network within a specific closed environment. In addition to this phenomenon has been the academic movement of schools since the end of the 20th Century, leading to the recruitment of university style research lecturers (doctors), where many teachers were former company engineers and who entered the teaching profession via different paths. The courses designed by subject matter specialists often lead to academic approaches to knowledge that are structured in a theoretical way. Faced with these two phenomena, other issues such as the development of market logic, the internationalization of companies and new forms of engineering led to contradictions. Innovation training aims to overcome these contradictions by creating methods that connect students to socioeconomic environments. It may refer, within disciplinary fields, to making students work on the basis of requests from companies through design offices or partnership projects. These methods can also take place outside school hours, such as the creation of junior enterprises or participation in technological competitions organized by major groups or associations, for example for the design of terrestrial drones and submarine robots. Immersion methods in companies also contribute to this openness to professional practices in the field, allowing students to tackle concrete problems.

A third dominant theme is that of the development of active teaching methods. This involves problem- and project-based experimental workshops, simulations and, more generally, collaborative approaches in which students are led to produce objects, concepts and answers to complex questions. Active methods have long been developed with the desire to break away from traditional teaching (lectures) centered on the transmission of inherited scientific knowledge. They are part of a socioconstructivist epistemology, according to which students learn better or differently if they produce or implement knowledge, through practical activities. Innovation training, which aims to develop students' capacities for initiative, intellectual speculation, conceptualization and experimentation, thus favors active methods. Classic teaching methods can offer an idea of what cultural and technical innovation represent, but the transmission of the capacity to innovate requires practice. Engineering schools have a long tradition of active methods, which they reinvent regularly as curricula become more and more academic. In addition to problem- and project-based learning, simulations and, more generally, scenarios also emerge from business methods, such as agile (customer-oriented and responsive) methods, or the application of specific procedures such as the Triz method (theory of solving inventive problems) [SON 07]. These teaching methods have their place in the new environments created to foster innovation. This is particularly the case of the *FabLabs*, originally created by the Massachusetts Institute of Technology (MIT), which French schools are now equipped with to provide students with tools for the design and manufacture of objects. This is also the case for *living labs* in which schools sometimes involve engineering students.

Innovation training therefore involves a wide variety of methods around common principles that bring them together. However, the curricular orientations chosen by schools show that choices are sometimes quite different in terms of pedagogical objectives.

Thus, we observe methods geared to the specific control of companies, alongside others geared to broader societal needs, leaving students with more freedom to make choices (such as support for people with disabilities). Some methods are oriented toward the concrete production of technical solutions to a circumscribed problem, while others seek to trigger learning, independently of an actual production. Some methods are limited to a specific scientific field, while others are intended to be much broader, mixing, for example, technical, economic, social and environmental issues. Hence, there are distinct differences and tensions between different conceptions of what engineering innovation training means in schools.

In order to better understand what is being played out in terms of the educational goals within pedagogical methods, the *Innov'Ing 2020* project survey also focused on the study of three specific innovation training methods developed by leading schools on the subject. The objective is to identify the educational logic that animates pedagogical practice.

10.3. Three innovation training methods

10.3.1. *Case studies of three engineering schools*

Understanding how higher education methods train engineers to innovate motivates us to study some of these methods in a concrete way. The study presented here aims to describe, observe and analyze the link between pedagogical methods and innovation training.

Common criteria choices have been formulated to allow a comparison between the methods. The common points relate to the criteria that condition any innovation training mentioned above, that is to say the decompartmentalization of discipline, openness to socioprofessional reference practices and active pedagogy. The selected schools do indeed organize interdisciplinary courses, are open to the socioeconomic environment of engineering education, research and internationalization, and also implement collaborative approaches per project.

In addition, differentiation criteria were retained concerning the geographical location, the predominant industrial and technological sector, the engineering sciences taught, the academic disciplines involved and the pedagogical choices made. Beyond the analysis of concrete cases, the objective is to identify certain invariants, or even to model innovation training in terms of their approaches and practices.

The survey carried out within the *Innov'Ing 2020* project identified three engineering courses throughout metropolitan France. Selected schools claim some form of prior expertise in their educational innovation training practices and are recognized as such by their peers. The different higher education institutions training engineers in France all have different statuses: schools under the supervision of the Ministry of Higher Education, schools under the supervision of the Ministry of Industry and schools that are integrated in universities. The engineering degrees they offer such as Masters of Science and Engineering CTI-certified and internationally recognized by the accreditation agencies. The innovation training methods proposed in the three cases were designed for and implemented in an entire student year. These methods are built on a sociotechnical approach to training, displaying societal and ethical values. Projects seek not only to design new uses, but also to respond to a social demand. Their organization is linked to their vocation to innovation, because teaching and research are conducted in project mode in a transversal and interdisciplinary way. It also targets proximity between the different actors of the school: management, teachers, researchers, support services and students.

The various training methods studied were the subject of direct observations by students, documentary studies and analyses of the school's website. Finally, interviews were conducted with the school head, training director and the person in charge of the method studied.

10.3.2. *Presentation of the three schools*

The innovation training methods of the three institutions are presented according to their institutional structure, their conception of innovation and the characteristics of the method. They appear in chronological order of their creation.

The first institution (School A) was created in the 1980s for the purpose of international openness. Since then, it has been integrated into a newly formed group of engineering schools under the supervision of the Ministry of Industry and the Digital Sector. Its teaching and research fields cover mainly telecommunications and electronics applied to different sectors of the industry. Innovation is defined as "the design of new products and services with a social need and the need for use". The purpose of the proposed "innovation training" is to develop "innovation skills" in order to invent new components, materials or processes for the design of new innovative products and services. The training content offered includes an introduction to research, reasoning, the human and social sciences as well as a reflection on the exact sciences and the human as well as social sciences underpinned by global systemic thinking. Traditional formats of lectures, tutorials and projects are also offered in the school. School A's innovation training is based on a project system. The project selected for the survey concerns all of the fourth semester students, constituted in teams of five who deal, for example, with problems such as the development of a system of geographical maps, the management of geolocalized information or voice recognition for connected homes. Each team responds, in the short term, to the request of a company reformulated by the researchers, so that the terms of the project are realistically achievable over the semester and compatible with future school research projects. The students, as subcontractors, see the results of their study and design projects validated by the representatives of the companies and evaluated during the coaching by the teachers.

The more recent second school (School B) established in the 1990s is a school integrated in a university under the supervision of the Ministry of Higher Education, Research and Innovation (MESRI). Upon creation, the school developed a strategy centered on the niche of innovation and decided to organize its training departments according to transversal and interdisciplinary project modes. This school is generalist, with specializations in materials engineering, mechanical and energy engineering, systems engineering and industrial engineering. The school approaches innovation upstream and with a broad perspective, defining it as an "idea generation process for creating new technological and non-technological objects that meet a market and offer added value for the enterprise in its environment". Innovation training is organized into short and medium-term projects spread over the 3-year curriculum with a systemic innovation objective that aims to develop "innovation engineering skills" to audit and model existing innovation processes to improve or create new ones. Innovation in products, services and processes can generate organizational, social and societal innovations. The method chosen for School B concerns a 2-day project in which student engineers will, with the help of an idea generating process, which will remain confidential and under the ownership of the school, propose to partner companies one or more technological or non-technological objects by ending the design at the model or prototype stage. Here are some examples of topics discussed by students: an Internet application that facilitates comfort, health and well-being in the home, a meal of the future with 3D printing, the factory of the future, new vehicle access functions or innovative solutions around construction. This project approach is characterized by the school as "innovation management" that, in addition to a systemic approach to creativity, relies on students' learning abilities such as self-study, "learning to learn" and learning as part of a team, in connection with a reflexive approach.

The third institution (School C), established in 2010, is also an engineering school, part of a university under the supervision of the Ministry of Higher Education, Research and Innovation (MESRI) . Since its creation, the school has been dedicated to innovation, which has a direct impact on the transversal and interdisciplinary organization of its courses. This school offers four specializations: IT, telecommunications and networks, ICT and health engineering. The school sees innovation as "the result of constructing a representation of the complexity to act, decide, identify and characterize the interdependencies between society and sustainable development, or between technology and the environment (functions/environments), in order to create new uses. Training through innovation is about developing skills that aim to "create a world of innovation, creation and technological challenges". The training allows innovative projects to be managed in their various dimensions: products, markets and technological systems in a societal context. These projects are the subject of four themes: politics, society, technology and management, designed to imagine new solutions to existing or emerging issues. The training contents are organized in relation to the innovation program, where the humanities play a dominant role through the teachings of subjects such as the study of the history of technologies or the philosophy of ideas which evoke, for example, the structure of technological revolutions. Innovation is also present in the human sciences through management courses or expressive practices in technological and non-technological contexts. The system chosen for School C concerns the transversal and trans-semester project led by the same team of engineering students who structure the innovation training course during the 3 years course itself. An unfinished project can be passed on to a team of students from the subsequent semester. The projects concern, for example, the ergonomics of a smartphone interface for mobility, supervision software for remote work management, autonomous vehicles in an environment closed to urban traffic or a prototype of a fair trade e-commerce website for farmers. Each team engaged in an "innovation engineering" process builds an innovation offer independently of any need that could have been formulated by a company. Only after some form of project maturation can the student teams identify one or more potential partners, including companies, with whom they can enter into an alliance in order to lead to a completion of the project.

10.3.3. *Three training methods dedicated to innovation*

The following three cases explore the innovative dimensions of the methods presented as such by the engineering schools studied. The first method concerns the development of a technological application. The second presents a process of generating ideas for the creation of a technological system and a non-technological process. The third method describes an engineering approach to designing a new technology for a new social purpose.

The project described as "innovative engineering project", following the School A's method, has two major objectives: to devise the development of a product or service with an innovative technological dimension and to improve students' performance in written and oral communication. Here, we take the example of a project carried out by four students over the course of a semester, which illustrates the method. A company, solicited by the research lecturers, added new features to its home automation voice recognition box to perfect it.

Adding features requires a new browser interface that the student team had to complete. After having understood and reformulated the requirement of the corporate sponsor, the students proposed a methodology by dividing the work team into three divisions: voice recognition, identification of the source of speech and implementation of the requests for the new interface.

Beyond their proposal of a global prototype including diagrams and the development and the integration of the prototype, the students had the initiative to add an algorithm to detect the source of speech in order to offer the customer new business opportunities for the home automation box. The project's school forum, open to all stakeholders, whether companies, institutional partners, researchers, peers and school staff, made this possible, using a feasibility study, demonstrations, a model and a poster, to communicate on innovative pedagogical modalities new to students. Using the project approach linking engineering sciences and management illustrates, through the example of proposing an algorithm leading to a new feature for the customer, how an innovative training approach can contribute to innovation both technologically and commercially.

The innovative dimension of method B lies in the implementation of a design-based training on an idea generation process, owned by the school, which provides the methods, tools and techniques to create new objects, in a broad sense, in the areas of technological and societal innovation, process innovation, processes and new products. Using process, students, coached by teachers and supervised by company representatives, worked for 2 days as part of a creative international collaborative operation that brought together partners, institutions, companies and organizations from different backgrounds. The following example demonstrates the training's innovative nature from two perspectives. At a first operational level, a team, collaborating with a partner company on themes concerning society and the aging population, proposed first a concept of a multiuse access control system in a dependent person's home. At a second level, the team, engaged in a reflexive process concerning its own functioning, proposed to adapt a personality test tool, used at the very beginning of the project, in order to analyze the team learning process while carrying out the project. The process of generating ideas, mobilizing creative techniques such as divergent thinking, the association of ideas, analogy, the dream-based method or different heuristic processes can thus raise issues and generate very different concepts within teams of international and multicultural students in this way fostering the emergence of synergies and cross-cultural skills. The students became involved by mobilizing the school's values that engage their reflection on the human and social impact of their activities. They managed the project's sessions in its various phases of collective creativity until the prototype phase, therefore testing their ability to manage innovative projects in a complex environment.

The latest innovative training method proposed by School C is illustrated by the example of a project devoted to the design of an e-platform to help dependent people with their daily food intake, taking into account meal composition, the management of purchases and the optimization of food resources. A team of students in Information Science and Technology (STIC) and building engineering, in the context of the project, generated several ideas around home-based products and services for dependent people. The design process required research, sources and contacts with personal care services, social agencies, hospital health professionals and nutritionists. Students' reflections on the economic, political, social and ecological environment were inspired by ethics courses.

Thus, the project's design process is both motivated by lessons on the history of technological revolutions and the philosophy of ideas, overlapping different environments and weaving links between different technological and non-technological fields. The design and innovation process also involved exploration and co-construction processes, individual and collective reflection, which called for a certain ability to think outside the box. The triangulation of a complex environment, interdisciplinary teachings and a process of building scientific knowledge within the project, allowed students to develop skills in responsibility, autonomy of action and thought, initiative and creativity required for innovation training. All of the pedagogical methods deployed here, during the project, represent the culmination of the creation of a higher education institute initiated by an innovative training strategy structuring the entire engineering cycle around a 3-year project that exposes students to the complexity of reality. The innovative nature of the training consists of relying on the plurality of approaches and the interdisciplinarity of engineering sciences teachers involved in a project led by the human and social sciences teams (SHS).

10.4. Innovation training teaching methods and logic

10.4.1. *Three divergent approaches to innovation training*

The first example of a method (school A) for "technological development training" aims to develop technical skills, and in the mid-term, products and services to meet the companies' specific needs. Companies are contacted by teachers who design and manage the entire project and its modalities. Technological development is a response to the precise order of a company that takes on the role of sponsor, and which has already given detailed instructions for a result that it already knows, for an application that must be operational in the short term. The skills mobilized by the students are the application of scientific theories (material sciences) that do not really involve a research approach. It is the same for the human sciences as it is mainly a matter of applying business management knowledge in this case. Students work reactively and have little autonomy for conducting the project, but can, like in the example, take innovative initiatives. It appears here that pedagogical modality is indeed innovative for students who have never done project-based learning, but it is perhaps less innovative when it comes to the method itself, the school or the company. The project mode does not necessarily induce innovation.

The second example (School B) of training in sociotechnical creativity aims to structure a process of generating ideas in order to create, in the mid-term, new technological and non-technological objects. It is primarily the teachers who control the exploitationist-type training [CROS 07] when they deliver the idea generation process used by the students in order to express innovative proposals. Companies act as partners and advisers and in doing so formulate a need, participate in the collaborative process and are an integral part of the project because they need to propose ideas themselves.

The generation of ideas through a collaborative activity over a constrained period of time makes use of the students' skills to manage, autonomously for 2 days, the process of collective creativity. In this dynamic, they rely on the humanities and social sciences and on teachers and companies that take on coaching roles. Beyond a process characterized by immediacy and spontaneity, students are expected to interpret the current and future needs of society.

The last example (School C) of "Innovation Engineering Training" is based on a representation of complexity and restores the interdependencies between technology, business, environment and society. Teachers, taking on the role of consultant experts, implement a framework in which students, with a great deal of autonomy, develop concepts that create changes in direction and bring existing or future technologies in line with the evolving needs of society, to create new applications. Their objective is to formulate, over time, an offer of innovation to recipients who are not identified at the start of the project. Student teams are engaged in advance and decide, in due course, to rally partners for co-development and implementation in the medium or long term. The skills that are mobilized are reflection, the ability to think outside the box and anticipation and interpretation of social needs according to sustainable development. This type of project can lead to a research approach with the aim of creating scientific knowledge.

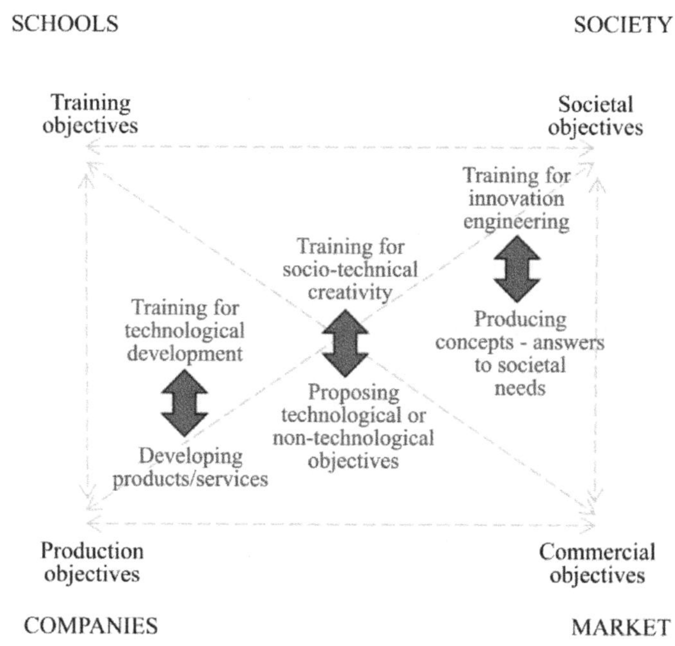

Figure 10.1. *Three types of training methods for innovation in engineering schools*

The three methods presented have distinct training objectives. Technological development, sociotechnical creativity and innovation engineering maintain different types of interest, nature and intensity with the objectives of production, marketing and sustainable development of a society as a whole. These different objectives can be, at the same time, if we take the perspective of actors, partners or recipients, in opposition to each other while also being complementary. Designing an innovation training method for engineers requires combining the more or less conscious and assumed expectations and needs of schools, businesses, markets and society. Developing products or services, proposing technological or non-technological objects and producing concepts in response to societal needs, use different content and skills, which each time raise the question of the pedagogical methods to be suitably associated with these innovative activities, if they indeed must lead to innovation.

The presentation and summarization of the three methods are not intended here to assess the relevance or not of a model that they could represent. As in the diagram, which tries to illustrate this aforementioned information graphically, here it is rather a question of organizing some reference points to facilitate the representation of what innovation training can be.

10.4.2. The logic at work in innovation training

The study of the forms of educational methods dedicated to the training of engineers shows us that tensions may arise due to the objectives and practices. Underlying logic emerges owing to certain constants (the relationship with entrepreneurship, the aims of innovation, technical knowledge, etc.), which are interspersed throughout the educational approaches. Beyond the curricular orientations and the types of approaches studied during the survey, there are in fact three major rationales that structure teaching practices dedicated to innovation training.

First, innovation can be understood as technological development, in the sense of the production of objects (real or digital), the development of new processes or the structuring of new services, which can be put on the market or provided to companies or customers. This type of innovation dominates training methods. It is visible in the first approach studied, and partly overlaps with the other two. All engineering schools establish methods dedicated to this type of innovation. The design and resolution of technical problems are at the heart of engineering and all students destined for this job must develop expertise in the field. Schools therefore set up, in the curricula, project plans and project and design workshops dedicated to acquiring this specific knowledge. During the last 2 years of study, at Master 1 and Master 2 levels, after basic scientific training, schools set up application-type activities in connection with implicated companies or professional institutions. Often, industrialists propose subjects, or delegate representatives in school, to oversee projects or study and design work. The creations can be more or less finalized. In some cases, large projects, carried out over long periods and in a collaborative way, can lead to manufacturing, marketing, concepts or exploitable results. When school research and development teams enter in contracts with companies, students sometimes participate in the work done hand- in- hand with their teachers. This is particularly the case for PhD students in engineering schools. In some configurations, the projects carried out by the students can lead to the creation of companies or the development of new structures within existing companies, according to entrepreneurial approaches, as valued by the CTI. This conception of innovation establishes scientific and technical knowledge as operational: dedicated to concrete applications and to economic and industrial performance. We seek to develop the students' capacity to use design or calculation tools to search for practical solutions and to achieve a technical result based on understanding the companies' needs. The dominant criteria are market driven, related to specific industrial sectors corresponding to job opportunities.

Another approach to innovation, partially visible in the three cases studied above, is research driven. Historically, French engineering schools placed little importance on research. Only a matter of a few decades ago, apart from in some major Parisian schools and university-type institutions, engineering schools possessed neither research laboratories nor researchers. With the internationalization of higher education [LAZ 99], schools have experienced a boom in academic education, intended to keep up with international standards that value research. Nearly all schools, have, in a short time developed research laboratories and recruited research lecturers (doctors). This movement has profoundly changed the curricular orientations of schools. Research enriches teaching content and working methods. As a result, students are frequently associated with the research activities of their research lecturers. The projects they lead are then quite different from the innovation conceived as technological development. The client, who is both the originator and the beneficiary, is more supportive of research laboratories than companies. This encompasses advancing knowledge from experiments, calculations, modeling and conceptualizations; more important than applicability is the production and contribution of new theoretical knowledge. As mentioned beforehand, scientific requirements often conflict with market requirements. Schools are seen as sources of conflict or rivalry between researchers specializing in industrial applications and development and researchers with a more academic speculative profile oriented toward experimentation and the academic valorization of research, particularly through publications. Scientific innovation (concept-oriented) is thus of a different nature to technological (product-oriented) innovation, both of which give rise to rather different pedagogical modalities, even if they can be combined in certain ways.

Another, more encompassing, concept of innovation is directed, not toward the market or science, but toward the needs of society and citizens in general. The third approach studied above gives us an example of this type of initiative. The project's reasoning is different from the other two insofar as the need for innovation does not originate from industrial companies or research laboratories, but from needs that are observable among the population. In this approach, it is often the students who build their own subject, develop the ability to observe their societal environment, identify problems, question the people concerned about their uses and their needs, thus constituting a central educational objective. This can be a response to real-life problems encountered by people in difficulty (the elderly or disabled), to propose improvements to our daily systems for sustainable development, or interventions in countries facing difficulties with populations lacking technical means. From a pedagogical point of view, the purpose of innovation is more general and can encompass the two previous approaches, insofar as it requires the mobilization of concepts and the production of objects or services. It does not necessarily fit into a technologically based logic that mainly consists of seeking the most advanced solution. Obtaining the most effective solution is the ultimate goal. This, indeed, can be less expensive and it simply involves the adaptation of existing technical tools. A differentiating point is also the mobilization of human and social sciences in this type of approach. They appear here not as support (such as the contributions of management to understanding market logic, or the contribution of communication, rhetoric and writing skills for theses) but as necessary theoretical contributions to projects to analyze the needs, uses and social acceptability of new objects. The ethical dimension of innovation is given greater consideration. This approach responds to more all-encompassing sociotechnical logic in that it combines both the social dimension and the technical dimension of a problem.

Figure 10.2 summarizes the three approaches of innovation training, according to which pedagogical methods are concretely implemented within schools:

Figure 10.2. *Three training approaches for innovation*

Three approaches are to be considered not only as dominant, but also categories of training objectives, thus making a combination of principles possible within pedagogical practices. The pedagogical tools available to the students can also fulfill two of these approaches at the same time. The approach of global engineering training encompasses the other two approaches (hence its position at the top of the triangle); insofar it combines them with the aim of achieving a broader goal on a societal level, and a more extensive mobilization of the students' knowledge. Certain constants reappear in the three innovation training approaches, such as tool creation, marketing relations technical conception and the role of human and social sciences. However, these different constants have quite different functions depending on whether the teaching method tends toward one or the other of these approaches. This diagram thus helps us to understand the intended objectives and the educational tools utilized in the different training approaches. Training objectives corresponding to these approaches all have their place at different intervals in engineering curricula. Some technological development projects occur during the study time dedicated to a specific discipline during a semester, whereas a humanitarian project can run transversally throughout another semester. The curricula proposed by engineering schools most often combine these three approaches, often giving greater focus to one in particular.

10.4.3. The challenges of a global approach to innovation

In general, the implication of human and social sciences in innovation training methods constitutes a key point in innovation training. In engineering schools, human and social sciences play a specific role compared to other disciplines [ROB 14]. Human and social sciences have allowed for a certain reflexivity within the training due the history of engineering, epistemology and the sociology of industrial sciences. As mentioned above, an engineering school curriculum gathers an array of disciplinary knowledge, albeit sometimes at different extremes. The coherence of the curriculum can only be understood by comprehending the engineer's role in companies and in society, of the evolving needs of humanity, of the place of technology, of the dominant ideologies concerning the market and the production of goods, and of the contemporary conception of innovation or progress in global history. The humanities and social sciences provide the necessary tools for understanding the aims of the training undertaken as well as the role of various disciplines in the curriculum, thus constituting a cohesive link.

The implementation of the human and social sciences differs within schools [LEM 03]. This can be seen as a tool to encourage the socioprofessional adaptation of students, through communication, supervision (team management, human relations at work) and management (project management, business economics). It can also be seen as an aid to enhancing students' self-development, the development of their creativity, their self-awareness, self-reflection, motivation, skills and professionalism. Finally, it can be seen as a means of gaining reflexivity that increases global understanding as well as incorporating the engineer's activities in a more general social, ethical and environmental logic. In these different approaches, disciplinary content and pedagogical modalities vary considerably.

In innovation training, the implication of the human and social sciences is relative to the specific objectives and desired outcomes intended by the curricula designers. In a certain way, the more the innovation is related to short-term objectives or limited to technical fields, the human and social sciences course content becomes less relevant. Indeed, whether it is a question of training for technological innovation via an application provided to a company, or training in the innovation of concepts or methods in a scientific field, the contribution of these disciplines is negligible. The more the innovation takes into account societal (long-term) goals, the more encompassing and mobilizing the approach. Their contributions concern both the tools of global understanding (sociological or anthropological surveys via questionnaires, interviews, observations) and the available theories and the authors' thinking. The evolution of society leads to a clearer integration of human and social sciences into engineering curricula to meet the contemporary challenges of innovation. Whatever approach is given to current problems (the ethical stakes, the respect of the environment, the acceptability of new technological products and the uses made of them), these cannot be ignored by engineers even if the objective is simply to adapt to the market, regardless of broader societal goals. This is why engineering schools, by already taking into account the logic of technological innovation and research innovation, tend to develop training systems geared toward global innovation and so integrate the technical, economic, ethical and social paradigms according to trans-disciplinary logic.

10.5. Conclusion

The study of the curricula of French engineering schools, carried out here on a broad scale drawing upon available literature, then in a targeted way by means of three case studies, allows us to discover the presence of different innovation training logics. These logics, when at work, are based on sometimes divergent objectives and disciplinary content, for example the divergence between the design of market-based products and the design of social solutions for the vulnerable, between the mobilization of human and social sciences as tools of communication or as tools for understanding social environments. These divergences bring to light conflicts in the adopted logic, thus leading to well-considered choices made by curricula designers (teachers or training managers).

Is there any logic preferable to others? Probably not, as in the end they finally appear more complementary than competitive. It is also observed that schools that implement innovation training in a comprehensive and integrative way, with a broad sociotechnical approach, do not abolish the training in technical innovation centered around targeted applications within academic disciplines. These different approaches correspond to different objectives and levels of training. In a curriculum that, step-by-step, leads students toward their degree, the application of technical knowledge, market adaptation, the response to company requirements and the anticipation of societal needs are all objectives dealt with throughout the curriculum, and which require progression and possibly different activities and practices. These training logics for innovation, both complementary and necessary, are intended to fit together. Thus, the projects that the students lead, in the aforementioned case study C studied above, implement the principles of technical innovation applied to the market, within wider social issues, in search of solutions integrating technical, economic and social aspects. It is therefore wishful that the schools, with regard to their specializations and their positioning in the engineering training landscape, will endeavor to integrate these different approaches in the best possible way.

Issues appear behind the pedagogical practices in terms of the engineer's role in society and the aims of technical innovation. Educational innovation at the service of innovation training is not just a question of technique, but a social practice that engages political conceptions and an ethical relationship with the world as well as the training methods proposed to students, they are expected to behave in a certain way, to enhance their problem-based analysis skills to a certain degree. This perspective therefore encourages curriculum developers to consider the training as an offer with regard to short-term objectives as well as more long-term social goals.

10.6. References

[CRO 07] CROS F., *L'agir innovationnel. Entre créativité et formation*, De Boeck & Larcier, Paris-Brussels, 2007.

[CTI 17] CTI, Références et orientations, Report, 2016, available at: https://www.cti-commission.fr/fonds-documentaire/.

[FOR 11] FOREST J., CHOUTEAU M., NGUYEN C., "Conceptions de l'innovation et formations de l'ingénieur", *Les cahiers du Musée des confluences*, vol. 7, pp. 37–47, 2011.

[LAZ 99] LAZUECH G., *L'exception française. Le modèle des grandes écoles à l'épreuve de la mondialisation*, PUR, Rennes, 1999.

[LEM 03] LEMAÎTRE D., *La formation humaine des ingénieurs*, PUF, Paris, 2003.

[LEM 07] LEMAÎTRE D. (ed.), "La formation des ingénieurs en France: entre logiques académiques et logiques professionnelles", *Les sciences de l'éducation. Pour l'ère nouvelle*, vol. 40, no. 3, pp. 7–96, 2007.

[ROB 14] ROBY C., Place et fonction des SHS dans les Écoles d'ingénieurs en France: état des lieux, enjeux et perspectives épistémiques, PhD Thesis, université Rennes 2, 2014.

[WHI 29] WHITEHEAD A. (ed.), "Universities and their functions", *The Aims of Education*, The Free Press, New York, pp. 91–101, 1929.

11

Developing Methods and Programs for Teaching Innovation to Engineers: Toward Eco-Innovation?

11.1. Introduction

The study presented in this chapter aims to contribute to the development of new programs capable of training engineers for innovation via an original and pragmatic approach applicable in various institutional contexts. This approach is the result of research conducted as part of the project Innov'ing 2020. This work demonstrated the need to revise teaching methods and to produce original pedagogical approaches that include a global or sociotechnical point of view in order to prepare innovators to respond to the social concerns of tomorrow.

The approach to innovation adopted here is holistic, aiming to link the technical dimension of innovation to its social, cultural, economic, political and ethical dimensions. It employs an increased mobilization of the human and social sciences (hereafter designated by the acronym HSS) relative to current curricula (see Chapter 10), and employs a more reflexive approach to engineering education favoring distance taking, critical thinking and the encouragement of the kind of awareness of social surroundings necessary for producing situated innovations (see Chapter 1).

Chapter written by Catherine ADAM and Serge COCO.

What types of educational methods meet these expectations? How can we design new courses better adapted to the needs that we identify? How can we train engineers for sociotechnical or socially situated innovation?

Rethinking innovation training in a pragmatic and ameliorative way does not mean that we must discard past pedagogical practices, but rather that we must start from the analysis of existing practices in order to design training methods that can, to take a naturalistic metaphor, take root, develop harmoniously and be grafted onto different training contexts. In this perspective, this chapter proposes an approach to designing innovation training adapted and transferable to multiple contexts, resulting from pedagogical research. In the first part, we present some basic theoretical and methodological elements that can be taken as frames of reference. Then we will explain the design methodology behind this pedagogical model based upon the analysis of the pedagogical practices employed in training engineering students at a French engineering school. Finally, starting from this fieldwork, we will illustrate how to put into place a flexible training method based upon the notion of the innovation fostering ecosystem.

11.2. A conception of sociotechnical innovation education

11.2.1. *A holistic approach*

Our proposal is based first and foremost on a comprehensive and inclusive vision of innovation training, approached holistically: from parts to whole, based on a complex approach to systems in which "everything is both more and less than the sum of the parts" [MOR 90], and in which each interdependent part participates in the whole. The concept of the pedagogical "tool box", elaborated from a sociotechnical perspective [ALB 10], and thought of as part of an overarching pedagogical system, is at the core of our approach and provides the conceptual basis for our entire methodology.

In recent years, literature on engineering education attests to the use of this type of pedagogical approach. In 1995, some authors in the United States were already calling for the development of integrative and holistic engineering education [BOR 95]. By suggesting a change in intellectual culture, they hoped for a more holistic approach in which processes and knowledge would be woven throughout the program and through which the focus would be on developing human potential with a new pedagogical approach in the face of contemporary challenges. In schools, students would then benefit from an intellectual environment where they would acquire skills in synthesis and analysis, gaining an awareness of the impact of technology and innovation, as well as an appreciation of engineering as part of an integral process of societal change, with a capacity to acknowledge responsibility for the progress of civilization [BOR 95].

Several authors have also emphasized the need for hybridization in engineering education that would allow learners to solve technical problems, not only using technical skills, but also transversal skills, especially analysis, to enable them to understand and to place these problems in their global environment, with a particular focus on their socioeconomic environment and on the process of learning and developing appropriate and sociosituated responses (see among others [AKA 03, JAM 11, KEN 05]). Adopting a sociotechnical approach means those innovations can and must come from diverse sources and that innovation is often situational and it very often results from encounters with different needs, diverse perspectives and multiple sources of knowledge. Scientific discoveries emerge out of scientific praxis, the complex series of simultaneous and multilevel interactions between socially situated actors and objects [PIC 10]. The same goes for the creation of technical objects, whether material or symbolic, which are at the heart of the process of designing educational and training methods for innovation presented here. The nature of technical objects should not be seen as secondary or incidental in the process, but they must also be understood as forming part of a larger ecosystem, which is indispensable for the development of innovations. This is another way of saying that even the innovations due to advanced scientific research cannot be understood in abstraction from their sociotechnical contexts, the meaning of society being broad enough to include not only human actors, but also the institutions within which they work, as well as the technical tools with which they work. Acknowledging the complexity of educational and professional developmental contexts thus require a shift in perspective that, as Albero [ALB 10] points out, "compels us to take into account the dependence relationships between activity, technique and environment" and leads "to more ecological systemic conceptions, in G. Bateson's sense of the term (1977) [...] focusing more on interactions between elements".

11.2.2. *A systemic conception: toward eco-innovation?*

Based on these epistemological reflections and previously obtained research results, the approach adopted here does not consist of the development of a simple pedagogical artifact, digital or not, like the digital training platforms, serious games, MOOC, which could be used identically in multiple contexts. In fact, engineering curricula already include examples of innovative pedagogical approaches: FabLabs, collaborative tools, entrepreneurial initiatives, etc. Without discrediting the relevance of this type innovation tools for achieving certain targeted objectives, as well as increasing the motivation of learners, we perceive the limits of these approaches based upon the uses that are made of them. It is a question of breaking with the idea of a direct causal link between the modernity of the tools used and the perceived innovativeness of practices that they permit [ALB 10]. For example, the deployment of digital teaching/learning platforms, modern in terms of computer development, sometimes excuses a traditional approach to teaching. This is manifestly the case where students watch a filmed lecture course, then answer questions about acquired knowledge via online quizzes. It is important to distinguish between learning objectives and the means used to achieve them. New objectives sometimes require the use of these new techniques, but using new means is not an end in itself.

Thus, more than the creation of innovative digital tools, our conception of innovation training corresponds to the development of training systems, envisioned as ecosystems including not only new tools, but tools adapted, and "situated" according to the educational environment, learning objectives, a desire to respond to both contingencies and to educational and societal expectations. We propose to call this systemic conception of innovation eco-innovation, with this neologism referring to both the terms of ecology and economy. Ecology refers to the set of relationships (intrinsic and interdependent) and interactions between different actors and their environment as a whole (social, economic, culture, etc.) out of which an innovation will be born, that it is to analyze and take into account.

Bourdieu's encourages us to view the economy within of a given ecological system in a broad sense [BOU 82], including the idea of economy of exchanges, practices and representations, with the notions of market and real or symbolic capital. By taking into account both the ecological and economic dimensions, the eco-innovative approach generates situated innovations (see below).

One of the main objectives of our pedagogical vision is then to prepare learners and future professionals to anticipate, analyze and act in complex sociotechnical situations in ways informed by their reflection and their awareness of the different issues and impacts (social, economic, political, ethical, etc.). Accomplishing this demands a transversal rethinking of curricula at all levels, and with respect to all involved actors. In this, our proposal is project-based teaching, in the broad sense[1], in which all the actors are involved, in a horizontal way, to the extent that the speech and the ideas of each enrich the global project.

11.3. Modeling a system for training innovators: an empirical-inductive approach

11.3.1. *A heuristic approach*

As we mentioned in section 11.1, the goal is to develop modular education systems that can be transposed and implemented in other current training contexts. In order to accomplish this, we have chosen as an object of study an existing French engineering school. This empirical-inductive approach allows us to define the different phases of this method in a heuristic and realistic way, and allows us to illustrate what this means in practice. Our objective is to provide a generalizable model that will allow others to appropriate this approach according to their current needs and to be able to transfer and situate it within their own teaching/learning context. Thus, in the following we identify and analyze the core elements in an already existing educational system, namely the 3-year foundation curriculum at a French National Engineering school, with the aim of developing a comprehensive and concrete pedagogy.

11.3.2. *A case study illustrating the need for interdisciplinarity*

When analyzing engineering training curricula, we notice that several transversal skills (being able to identify, analyze, problematize, mobilize, transfer, etc.) are developed independently via different areas of study.

1 Educational projects, but also school projects.

However, during trainer evaluations and assessments, difficulties of acquisition and/or transferability of these skills are regularly noted in other situations, academic or professional, including the ability to determine to trouble-shoot and recognize and develop problematics [FAB 06].

One of the points of view here is that innovation training cannot be defined as a single skill to be acquired by students through traditional pedagogy. The ability to innovate, from a sociotechnical point of view, should rather be considered as a sum of different and interdependent skills, capable of being acquired in different ways (academic and non-academic), at different levels of education over the course of life-long learning. In this sense, the ability to "innovate" is recognized to be a meta-competence encompassing various essential skills.

Using Bloom's taxonomy [BLO 56], we began by determining a list of cross-curricular competencies that constitute the "meta-competence of innovation" that would be acquired by the student, and which is in keeping with the present descriptions in the recommendations of the Commission of engineering studies in France[2] and the description of the acquired competences necessary to obtain an engineering degree, and which are likewise in agreement with the research results acquired by the Innov'Ing 2020 project.

Being able to innovate – a meta-competence	
Individual skills	Cooperative skills
Ability to	
Develop	Absorb
Diagnose	Communicate
Diverge	Compromise
Evaluate/assess	Coordinate
Identify	Adapt
Observe	Integrate
Optimize	Interpret
Organize	Lead
Troubleshoot	
Mobilize	
Transfer	

Table 11.1. *Being able to innovate: a meta-competence*

2 http://cti-commission.fr.

This non-exhaustive list includes skills related to personal development and others related to interactions with others. We then analyzed each course description of the training program studied, from the point of view of the proposed contents, starting from the postulate that these descriptions, without being necessarily the reflection of actual practices, could help us to perceive if these skills were made for the purpose of writing learning objectives formalized on paper.

Of these 19 competencies, 14 are present in the descriptions at different levels (see examples in Figures 11.1 and 11.2).

Course title: **Mathematics and Computer Science 2**	
Teaching area: Mathematics, Computer Science	
Name of supervisor:	
Semester(s): 2	ECTS credits: 5
Scheduled classes: 60 hours	Self-study time: 50 hours
Language: FR	
Common program: yes	Professional profiles: all profiles

CONTEXT AND SUMMARY DESCRIPTION

This course proposes both an introduction to the essential notions of probability and statistics, and the application of algorithmic concepts, especially over the course of a project.

EVALUATION MODES AND CRITERIA

Probabilities and statistics are evaluated based on a written report on class content, and a final individual written exam. The computer project will give rise to a report and an oral examination. Programming theory and dynamic data structures will be evaluated via a written test.

CONTRIBUTION TO THE ACQUISITION OF TRANSVERSAL SKILLS

This course contributes to the acquisition of skills in the field of scientific and technical culture. In addition, accomplishment of the project also requires the development of management and project administration skills.

Figure 11.1. *Mathematical and computer science course title*

Thus, after having identified these skills, we met with the department and program heads in order to explore how they approached these objectives and aims based on their discourse and their representations of these notions as well as their effective implementation of skill-based approaches in their curricula. During our interviews, which were semistructured, the teachers mentioned that at the beginning of the curriculum, an engineering student analyzes problems or requests in a fragmented way in order to recommend a technical solution. There is not yet a blueprint of an interdisciplinary solution and certainly no recognition of the fact that many specialties are interdependent upon one another. The engineering student is thus brought to solve problems in the context of a discipline or subject, without having been asked to develop a global vision, analysis, reflexivity or transfer of previously acquired skills.

Course title : **SPID Project – Implementation phase** (Systems, Perception, Information, Decision)	
Teaching Area: Electronic, Computer Science–Mathematics, Physics and Marine Environment Systems	
Name of supervisor:	
Semester(s): 4	ECTS Credits: 7
Scheduled classes: 120 hours	Self-study time: 20 hours
Language: FR	
Professional profiles: ROB, GL, PSO, TSE	

CONTEXT AND SUMMARY DESCRIPTION

The project conception phase and the project implementation phase are two courses which aim to develop the technical, methodological, human and organizational skills of students, to professionalize them and to lead them toward autonomy through guiding them to conceive and implement a technical project over two semesters in teams of 4 to 6.

The learning objectives of the course are:

– **how to work in a team:**
  - to compromise with each other's methods
  - to coordinate themselves
  - to integrate themselves within a group (participation, listening, respect)
  - to develop their ability to communicate and manage conflicts
  - to lead and to assume responsibility
  - to make decisions and take initiatives collectively
  - to make contact with resourceful people

Figure 11.2. *SPID project course*

In order to enable the student to analyze a complex situation, to understand and extract the relevant information, the engineering student needs, according to the teachers, to "make the connections". He must be able to establish bridges (transversalities) between one competence and another, mobilizing several in new situations (transferability). This is an example of the necessity of transitioning from the world of the academy to the "industrial" world that emerges in the teachers' discourse. Not focusing solely on the end result of the work done by the students, but rather on the creative process employed while completing the work can help to encourage the development of this capacity [BAR 11].

These analysis results confirm the need to establish transversalities between the disciplines, in other words to cultivate an interdisciplinary approach. It should be noted, for example, that one of the learning objectives announced to students of the SPID department project (see Figure 11.2) is to "connect the disciplines". However, this connection is framed as concerning separate technical disciplines and technical knowledge and social scientific knowledge.

Our analysis of the curriculum of this engineering school also revealed that the five skills identified as essential in the context of sociotechnical innovation training (being able (1) to diagnose, (2) to interpret, (3) to diverge, (4) to absorb, (5) to transfer, see Table 11.1) stand out in that they are not mentioned in any curricular description of the program being studied. For our part, we particularly paid attention to these skills and their absence from the explicit program, as this allowed us to establish the need for developing a complementary educational project that could adequately respond to student needs (see section 11.4.2).

11.4. The mobilization of HSS within an existing program

We carried out an experiment to design complementary methods based upon what we had seen, and in correspondence with testimonials gathered from the teachers. It was then a question of concretely presenting ways of working to design eco-innovative training tools. Thus, the difficulties and shortcomings identified, from our empirico-inductive analysis of our case study served as the empirical anchor points that allowed us to model and develop an adapted and situated teaching system.

From a holistic perspective, we then decided to develop a curriculum that addressed all levels of the educational ecosystem: the macrolevel (institution), the mesolevel (discipline) and the microlevel (classes/lessons).

The most ambitious element, and certainly the most difficult to implement within an already existing educational institution, is the macrolevel, since according to the "eco-innovative" approach, and in order to change perspective and develop a comprehensive sociotechnical approach, it would make sense to suggest that HSS be integrated into the totality of the engineering curriculum. Luckily, this is sometimes already the case in French engineering schools (see, among others, Chapter 10). At the mesolevel, we propose to put the development of transversalities at the heart of the global engineering education by opening up all of the disciplines to interdisciplinary approaches. Finally, at the microlevel, the educational tools would include specific and complementary training elements derived from HSS research.

Apart from the macrolevel, which would require other broader considerations in terms of educational and institutional policies that we cannot detail in this chapter, the proposed system for training innovators has two components that relate to applicability: on the one hand, the implementation of possible transversalities within the disciplines taught (section 11.4.1), and, on the other hand, the development of a specific, complementary and transversal training module, which will result in an increased mobilization of HSS training within the curriculum (section 11.4.2).

11.4.1. *Creating transversality by opening up the disciplines*

In our view, the prerequisite for holistic innovation training is the idea that it cannot be unilaterally confined to "innovation" training in a single class or a single department, but that it is necessary to think and conceive of it via a systemic approach and in a transversal way (see above).

The qualitative analysis of these educational tools and of the teachers' discourse shows that a number of courses propose the development of the key skills that we have identified in section 11.3. However, teachers point out not only difficulties in the acquisition, but also in the remobilization and transfer of these skills to other academic and/or professional contexts. These difficulties are therefore considered as points of entry, allowing us to conceive the first part of our method, that is, to establish transversalities between the hard sciences and the HSS.

In practice, the implementation of these transversalities to an interdisciplinary curriculum can be explained from the two teaching examples proposed in the curriculum studied: a department of mathematics and computer science proposed within the framework of the common core of the first year of the engineering curriculum (see Figure 11.1) and two specialized diplomas in systems in perception, information decision (hereafter designated as SPID) within the second year of the engineering curriculum (see Figure 11.2).

These two examples correspond to two different scenarios:

– in the context of the SPID Project, the second-year students must work together in teams of four to six people, they are challenged to draw on their area of expertise, with the aim of preparing them for jobs "at the heart of innovation"[3]. They must, therefore, create a group dynamic, organize themselves, communicate and manage difficulties with the participants' investments in time and energy, reliability (their ability to complete a job on time), know-how, etc. This work employs not only peer evaluation, but also a form of accompaniment similar to tutoring. In the context of this project, many of the skills identified above are mobilized ("compose", "coordinate", "develop", "communicate"). It is then up to the teacher to increase student reflexivity with activities that will stimulate the learners (intermediate assessments, co-evaluations, implementation practices that encourage the development of co-construction via training by peers, tutors, learners, trainers). We observe that transversalities or the pooling of intellectual resources could be established between this project and some lessons/teachings offered in HSS, such as courses on leadership or intercultural management;

– in the mathematics and computer science department, students had to carry out a computer project in pairs to develop specific computational

3 "The SPID option offers engineering students the opportunity to apprehend complex electronic and computer systems developed in numerous sectors of activity (robotics, telecommunications, transportation, aeronautics, space) in connection with new technologies. The skills acquired enable a wide variety of professions at the heart of innovation". Excerpt from the presentation of the SPID option made by the school studied.

concepts by integrating both required problem sets, and also by proposing and justifying individual solutions. Time management, commitment and investment in this work as determined by the teacher to be the conditions of success or failure. The "professional" goal, estimated as desirable by the teacher, is for future engineers to be able to program 100 lines of code. In this context, the term "to lead" does not have the same meaning as what we have previously discussed and does not involve the development of the capacity for reflexivity in the learner. What we are talking about here is a process of applying acquired knowledge. However, from the moment when the teacher expresses that, within the framework of this project, the following are also evaluated: "the work produced, the integration of obligatory elements and the justification of the choices made", we are well in the context of a situation which, in the engineering profession, potentially requires implementing innovation or design skills. It is in this type of gap that we envisage a possible accompaniment of the teacher toward the identification of the whole range of skills underlying innovation asked of the students within the framework of this project. Because of the contribution of HSS work, students will be able to apply a more systemic and "eco-innovative" approach toward exploring and resolving problems.

In summary, in order to better understand the design of this first part of our training, we have presented and analyzed two scenarios, which illustrate two possible implementations of transversalities in an interdisciplinary approach:

– either the skills identified by the project are in the direction of the skills allowing for the development of the capacity to innovate. The contribution of the HSS can then be situated as an accompaniment to heighten the reflexivity of the learners, to encourage them to analyze the processes implemented and to raise their awareness the skills that they are developing in order to be able to better transpose and remobilize them in the future. This work of accompaniment would then be accomplished in a partnership between the teachers of HSS and those of the technical disciplines;

– or the skills identified in the projects, analyzed in the light of the teachers' discourses, do not support sociotechnical innovation, so there is a need for upstream support for the project advisors, either to clarify expectations or to develop teaching/learning objectives and possible transversalities with HSS. Nevertheless, it would still be of interest identify these skills to learners via a partnership between teachers of HSS and those of the technical disciplines.

These two examples show how, in an easy and inexpensive way, through the opening of the disciplines and the collaboration of the teachers of technical disciplines and HSS, it is possible to make the students perceive that the disciplinary learning makes sense in a global ecosystem, which must then enable them to forge links between the different fields involved and expectations related to a given new problem.

11.4.2. *Mobilizing HSS to establish particular skills*

The experimental research conducted in this field has shown us that certain skills, considered as constituting meta-competences necessary for innovation, were not covered by the programs studied, and that greater openness to HSS could be a good way to remediate this lack. We have therefore devised concrete solutions to supplement these courses, in correspondence with the expectations formulated discursively by the teachers. This second part of our training system is based on the contributions of the HSS in the training of future eco-innovators, and it involves training modules that are transversal and complementary to the other lessons/teachings identified.

Within this module, several courses appeared strategically necessary to us as to allow for the acquisition of the targeted skills. These include the following:

– Being able to diagnose (1) and interpret (2): To target these skills, a sociohistorical approach to problems is relevant. Aimed at the analysis and deconstruction of learner's initial representations of innovation and innovators in a broad sense, and highlighting the representations of different academic and artistic disciplines in society throughout history, it serves to open up critical awareness of these various representations and their value. We here employ a pedagogy of "contrast, confrontation and meta-cognition" [BRU 08, p. 183]. It involves leading students to draw on preexisting knowledge to diagnose and interpret a situation in its overall context. This sociohistorical approach to innovation also helps engineering students to deconstruct stereotypes and to develop their critical thinking skills.

Sociology courses also have their place in the phases of discovery of needs by helping students to learn how to identify by socially situated observation problems and their relationships to their environments.

For example, through the action-oriented approach, students can be trained in research, and more specifically in activity analysis [BAR 11]. Learning simple techniques for research and analysis in the social sciences, with the help of the various teachers–researchers involved, offers the means to observe and analyze situations in their complexity.

– Being able to diverge (3): In a complementary sociolinguistic approach, a work on their own representations of the engineering profession and innovation offers the students the possibility to become self-aware and put themselves at a distance from these representations to "go further", in particular with respect to developing the capacity for divergent thinking. Reflexivity and self-awareness are indeed processes that are necessary in order to develop the capacity to innovate.

– Being able to absorb (4) and transfer (5): Finally, the mobilization of HSS teachings to foster a reflexive approach at each stage in the innovation process offers students the means to become aware of the steps necessary to fully integrate these new skills and develop a capacity to transfer the acquired skills into other contexts. This construction of the capacity to innovate, in a school context, should be recognized as involving the co-construction of knowledge, know-how and skills, through interaction between students, teachers and professionals involved in the educational project.

Training innovators to understand contexts, the expectations of actors, economic, environmental, geographical constraints, etc., is a major pedagogical objective. This understanding and awareness leads students to work on intellectual, ethical and practical obstacles to develop creative reasoning skills. For this purpose, cross-analysis, the debating ethical issues and the consideration of scientific controversies serve as privileged pedagogical modalities.

The approach is socioconstructivist insofar as it aims at the identification by observation of problems as socially and geographically situated in their environment, social issues common and relevant to students in their immediate environment.

In this perspective, the questions of teachers around the development of students with problematic skills, which we have also encountered in our teaching, have marked our reflections during this process of educational design.

A central place has been given to the reflection around the activity of questioning and developing problem-solving skills via an "adaptive method", which considers that "the subject acts and by acting constructs the situation and constitutes itself at the same time" [FAB 06]. This approach leads to considering problematization, in the sense that this term is used by Michel Fabre, as a meta-framework, a reference framework that is flexible and which allows students, future engineers, to acquire these skills, to question themselves in a non-closed framework, with little constraint for didactic purposes, by inducers of problematization [FAB 09], so that they can integrate these skills, remobilize them and transpose them later.

The aim is to propose an approach that is not purely procedural, but also aims at learning how to build problems, by exploring the data, the conditions and the frames of reference, be they economic, technical, but also historical, social, environmental, ethical, etc. [FAB 06, FAB 14]. In this sense, the HSS has a leading role to play in the implementation of interdisciplinary projects, as drivers of problematization [FAB 09], that introduce different perspectives and postures complementary to those employed in the technical disciplines.

With this in mind, the proposals made above for this second application component can also be integrated into a macroproject, in the form of a transversal training module, complementary with existing assignments and projects, such as those described above, and inserted into the curriculum. At each step, the project would require that the knowledge acquired in the different technical disciplines be mobilized alongside transversal skills with both aspects being anchored in a reflective approach encouraged by the contributions of HSS coursework.

In practice, in our case study, a project such as the one proposed in the SPID project can become a transversal project, within the framework of a training module for the duration of the studied curriculum (3 years), starting from a hermeneutical and historical approach to the initial representations of the engineering profession and innovation, which will result in a deconstruction of these representations that would then be followed by a phase of co-construction for innovation. These transversal projects encourage students to produce services or new objects based on real research and social requirements, allowing the development of a new generation of innovators.

What is important is not so much the solutions that students develop but rather the process of eco-innovation that allows learners, who will then become engineers, to apply this process in novel contexts, using their increased awareness to confront and resolve problems in all their complexity.

11.5. Conclusion

The results of the Innov'Ing 2020 research project allowed us to determine the conditions under which we can think and design a pedagogy for fostering sociotechnical innovation. Within this framework, we have revisited the design of various courses and projects with the aim of producing pedagogical approaches that foster the innovation and creativity necessary to meet expectations, including the uses and demands of society, by taking into account the particular contexts of each act of creation. Some higher education institutions in France (see Chapter 10) and internationally (see Chapter 1) already tend to include ethics and sustainability education in engineering programs. We also propose to contribute to these training developments by presenting the concrete bases for an eco-friendly approach to innovation training. It is a proposal based on another conception of innovation training, socially and geographically situated, not disconnected from the reality on the ground. According to the ambitions of a holistic approach, this chapter offers ways to design pedagogical programs for innovation, each of which can be grasped, depending on its context and its educational needs, as a basis for diverse and eco-situated pedagogical implementations. This will allow, hopefully, a profusion of newly situated innovations, set in motion by a new generation of eco-innovative innovators formed in the dedicated and now more innovative institutions of higher education.

11.6. References

[AKA 03] AKAY A., "The renaissance engineer: educating engineers in a post-9/11 world didactique", *European Journal of Engineering Education*, vol. 28, 2003, http://dx.doi.org/10.1080/0304379031000078979.

[ALB 10] ALBERO B., "Une approche sociotechnique des environnements de formation. Rationalité, modèles et principes d'action", *Education et didactique*, vol. 4, no. 1, pp. 1–17, 2010.

[BAR 11] BARBIER J.-M., *Vocabulaire d'analyse des activités*, PUF, Paris, 2011.

[BAT 77] BATESON G., *Vers une écologie de l'esprit*, Le Seuil, Paris, 1977.

[BLO 56] BLOOM B.S, *Taxonomy of Educational Objectives: The Classification of Educational Goals*, David McKay Publications, Philadelphia, 1956.

[BOR 95] BORDOGNA J., FROMM E., ERNST E., "An integrative and holistic engineering education", *Journal of Science Education and Technology*, vol. 4, no. 3, pp. 191–198, 1995.

[BOU 82] BOURDIEU P., *Ce que parler veut dire, l'économie des échanges linguistiques*, Fayard, Paris, 1982.

[BRU 08] BRUNER J., *L'éducation, entrée dans la culture*, 2nd ed., Retz, Paris, 2008.

[FAB 06] FABRE M., VELLAS E., *Situations de formation et problématisation?*, De Boeck Supérieur, Paris, 2006.

[FAB 09] FABRE M., MUSQUER A., "Comment aider l'élève à problématiser? Les inducteurs de problématisation", *Les sciences de l'éducation pour l'Ère Nouvelle*, vol. 42, no. 3, pp. 111–129, 2009.

[FAB 14] FABRE M., WEIL-BARAIS A., XYPAS C., *Les problèmes complexes flous en éducation*, De Boeck, Louvain-la-Neuve, 2014.

[JAM 11] JAMISON A., CHRISTENSEN S.H., BOTIN L., *The Hybrid Imagination. Science and Technology in Cultural Perspective*, Morgan & Claypool Publishers, San Rafael, 2011.

[MOR 90] MORIN E., *Introduction à la pensée complexe*, E.S.F, Paris, 1990.

[PIC 10] PICKERING A., *The Mangle of Practice: Time, Agency, and Science*, University of Chicago Press, London, 2010.

Conclusion

Perspective: Engineering Training, from Yesterday to Tomorrow

In our technological societies, the question of innovation is not new. Certainly, one can legitimately feel that we are living in a period in which the range of available products is infinite and where every day brings its share of novelties. No doubt we have entered a new phase of the Industrial Revolution with the gradual generalization of the digital world, which we feel that this ongoing process is already beginning to change our ways of life. Is such a phenomenon new? A look back at the past would remind us that our ancestors had to experience the same type of feeling. The invention of the steam engine and its development, concomitant with the exploitation of coal and the progress of metallurgy, allowed the development of the big industry, inducing new modes of production, but also transforming social relationships. One of the variations of this technological revolution was the creation of railways around the world that, in each country, remodeled land use planning and amplified the movement of the men and products. We use electricity today: pressing a button to light a room is a "natural" behavior and we pay no attention to it. But an interruption of the electrical flow over time such as we have known at the time of the great storms would paralyze a given region's economic and social life. Electrification, started at the end of the 19th Century, has definitively transformed the life of societies in a few decades, from industrial organization to the home, and we are the forgetful heirs: without electricity, no digital! At the same time, a new technological object was emerging that would radically change the relationship with space on earth, in the air and in the sea: the combustion engine, whose global reach

Conclusion written by André GRELON.

has not only generated an industry of rare power, but also geopolitical upheavals and wars to the present day, etc.

These technological innovations are due, in large part, to a set of individuals with a particular profile who have contributed over the centuries to their profession and have long been seeking social recognition. Unlike established and reputable professions, doctors, lawyers and clergy, who also represent a particular profession, engineers were by far the last to have formalized and widespread training. Since medieval times and the constitution of universities in Europe, the teaching of law, medicine and theology are codified, the steps in knowledge are controlled before being crossed and lead to the final recognition, namely, peer recognition, specific official rights and related social status. Nothing like the engineers who, if they received an education allowing them to read both in the vernacular language and in Latin, know how to write and draw and to have notions of geometry, also learned widely on the job, in training with more experienced engineers and following great projects like those of cathedrals[1]. For the individuals that the historians have identified, those who addressed the princes to make themselves known and receive an order, we know what expertise they exhibited in order to achieve war machines, fortifications, hydraulic works, automata, etc.[2] Perhaps this nomadic training, outside of scholastic knowledge and academic rigidities, encouraged them to develop their inventive genius.

When the first schools appeared, like the artillery schools in the Spanish Empire or in France, they only concerned a handful of technical officers. In the 18th Century, in a period of splendor for the arts, with the publication of encyclopedias devoted to arts and crafts, the foundation of schools of drawing, of machinery offices open to the public, craftsmen technicians designated themselves as engineers. They presented their inventions to the Academy of Sciences or the Office of Commerce and hoped for a distinction by a medal or a bonus: operators, manufacturers of various tools and machines, mill builders, they were different from the tradespeople due to their inventive capabilities[3]. Some even managed to obtain a title of engineer

1 [JAC 90].

2 [GIL 64].

3 [HIL 00].

of the king without passing a stamped training[4]. Across Europe, there were few official schools. In France, the country that counts the most, the two military schools, that of the Engineering located in the fortress of Mézières in the heart of the Ardennes and that of the shipbuilders based in Paris, produced, after a demanding training, technical officers in limited numbers[5]; the École des Ponts et Chaussées welcomes pupils in numbers, at least half of whom will not join the corps, but who will increase the population of civil engineers; as for the École nationale supérieure des mines that opened at the end of the last century, is inspired by the already famous École des mines de Freiberg in Saxony whose programs and innovative methods are attracting interest all over Europe. Its existence was short due to the lack of credit and the few trained technicians there struggled to find a job. These establishments maintained or restored during the Revolution, became application schools of the École polytechnique.

In 1829 in Paris, a handful of scientists and a businessman came together to open a new type of school to train industrial engineers or, to put it in their own words, "doctors from plants and factories". This was the École centrale des arts et manufactures, which is presumably the first institution to issue an engineering degree based exclusively on the completion of specialized training in 3 years[6]. To describe major innovation, we can say that for the promoters, it is about teaching the "science of the engineers", a new discipline, taking the sciences of mechanics, chemistry and physics by dedicating them to concrete applications. Soon after, the school welcomed students from many countries. The unprecedented model of this institution flourished: in Ghent, a school of civil engineering and in Liège, a school of

4 [VIR 17]: The man received training as a doctor at the université protestant d'Orange before leaving for Geneva where he would probably have trained in drawing techniques. Through the cartography publications and the compilation works that he multiplied, he managed to make himself known and obtain a title of engineer of the King, in the province of Languedoc, assigned a time to the Canal du Midi ... after having abjured his Protestant faith.

5 For this last school, its founder, Louis Duhamel du Monceau, was inspired by the graphical and mathematical methods used in England to build ships, upsetting the traditions that the master carpenters transmit from generation to generation, and that the cohorts of engineer builders are required to replace.

6 The École Polytechnique did not issue an engineering degree until 1937. Each application school sends its students once trained in its corps (Mines, Bridges, State Manufactures, etc.) in which they will climb the hierarchical ladder. Already members of the corps, as soon as they enter the application school, they do not need additional certification.

arts and manufactures and mines were founded[7]. Teachers at Centrale were previous students. Other alumni founded an engineering school in Lausanne in 1853 and the engineering school of Barcelona opened in 1851 and is openly inspired by the Parisian school[8], which is a clear sign that this type of organization and the curricula developed here corresponded to the expectations of potential employers and the economic needs at the time. Other institutions were created in Europe, be it in Lisbon or Prague, but engineers that were trained academically did not completely colonize the industrial world: they remained a small minority. Civil engineers existed, started companies, offered new products, built installations in private homes or in factories, and this population continued to grow over time. However, certification was not yet a requirement. From this point of view, in England a different model developed, which emerged in continental Europe. The engineers were trained in workshops, under the guidance of a paid "sponsor". They were recognized as such only when they were accepted into one of the engineering companies, the first of which, the Society of Civil Engineers, created in 1818. It is a society in which membership is permitted only at the end of a propaedeutic of several years[9].

It was necessary to wait until the end of the 19th Century and the beginning of the following century to see a process of generalization of certification in continental Europe, including in Great Britain. With the second industrialization of industrial electricity, electromechanics, organic and pharmaceutical chemistry, metallurgy of alloys, and other industrial disciplines, the need for specialized engineers was increasing. One of the pioneering establishments in this field was undoubtedly the Institut électrotechnique de Liège founded in 1883 in the impetus given by the first international exhibition of electricity in Paris in 1881, by the industrial patron Georges Montefiore-Levi and entrusted to the engineer Éric Gérard who, after studying engineering in Liège, received training at the École supérieure de télégraphie de Paris. Scientists and engineers from all over Europe came here to learn about industrial electricity, both theoretically and practically. Germany, which became the dominant industrial power in Europe, organized its technical higher education in eight dynamic and innovative *Technische Hochschulen* organized by specialized departments. In France, new institutions were opening, especially under the aegis of the

7 [BAU 93].

8 [PAQ 04].

9 [MAR 86].

faculties of science, which constructed, within the walls of the universities, finally created by the government of the Third Republic in 1896, institutes of applied sciences issuing engineering degrees that became the common title sanctioning higher technical studies. However, the country still had a number of self-taught engineers, to the point that in 1934 there was legislation protecting the titles of qualified engineers provided for the issuance of a particular title, that of a DPE engineer, given by the State, on the condition of providing a thesis justifying real engineering activities before a committee from the Conservatoire des arts et métiers. The records that have been kept highlight the quality of the work of these candidates and their often innovative nature.

It is starting from the second half of the 20th Century that the higher technological training programs were followed on an unprecedented scale. Thus, France, which was still largely rural in the inter-war period, became a major industrial power during the period known as the "Glorious thirties", and it was primarily the engineers who were entrusted with the task of transforming the country. Also, from the late 1950s, new schools opened (INSA, ENI), and this phenomenon of creation has not ceased since (technology universities, polytechnic university, etc.). This process common to our techno-scientific societies is correlative to the new industrial revolution, obviously marked by the exponential development of computer science and its derivatives, and also, among other things, by the rise of an innovative chemistry that fits into all sectors of production. At the same time, the development of trade and interstate trade agreements lead to the circulation of tangible and intangible products, leading to widespread international competition. Such a transformation impacted the whole of higher education leading to a "universitarization" of training. The curricula of the different countries have been forced to homogenize with the implementation of three levels of training called in France license, master and doctorate. This organization is supposed to be able to facilitate the movement of students across borders. As for the last noun, it implies that henceforth research is an integral part of a complete higher education program. Such a method will not be without impact in the mid or long term to the traditional structuring of French higher education, perceived as being constituted of the university world, on the one hand, and of what are known as the graduate engineering schools, on the other hand.

This detour by some past examples and the evocation of the role of education in the second half of the 20th Century lead us to consider the question of innovation as it arises today and the role played by engineers. This is what the 11 chapters of this book deal with. Devoting a book to career preparation for these professionals is necessary and logical, because the very function of engineers and their social responsibility, even beyond their professional obligations, that is to say that society expects of them, is to propose new solutions to the problems that emerge continuously in all sectors of economic and social life. This presupposes a vast body of fundamental knowledge, but also of proven know-how, because it is not only a question of promoting theoretical propositions but of achieving realizations, and these efficient realizations, what the common language designates under the name "skills". Their abilities must even lead them to anticipate the expectations, desires, if not the needs that are still unexpressed of all or part of the population.

It has been mentioned above that the engineer's goal since his origins, with or without academic training, has been to design and implement innovation. But today, this requirement seems to take on a new dimension, because the means of communication that we have make available to the knowledge of each, almost immediately, the latest inventions placed on the market. The vast process of circulation of objects in what is known as globalization provides the potential consumer with the most recent productions made on the side of the world. In a ruthless competitive environment, as customs barriers are erased, the imperative will be to offer something new, at the best price, and to make sure that it is the national companies that are the producers in order to keep jobs in the country and strengthen the nation's economic power. The insistence in institutions to train students to open up to innovation and to invest in it is described by the book's program, from diversified examples taken in France and in other countries that highlights the international nature of this requirement. Does this mean that school institutions have derogated from their historic mission and gradually settled into lethargic routines? It is undoubtedly a bad process. However, the fact analyzed in one of the chapters, that the *Commission des titres d'ingénieur* who, in France, oversees the programs of 206 engineering schools authorized by it, sends urgent recommendations to the training institutions and shows the national character of ardent obligation of this injunction. However, this notion of innovation has the characteristics of a portmanteau word. The definitions proposed in each of the chapters are only partially overlapping, certainly because the pedagogical teams have different conceptions of them, and also because they must probably take into account the relative specificity of each establishment and the industrial sector where the new graduates will preferentially make a career and the positions they will occupy in the companies where they will be hired. Basically, it is reassuring to see that there is no uniformity and that we have not returned to the period, fortunately far away, when the Minister of Education could congratulate himself that, each faculty, at the same time, the professors had to teach the same program which had been communicated to them by the central administration.

In addition, we can draw some general reflections on the reading of this solid and well-argued work. The contemporary engineer, as the current training wishes to model it, has seen an extension of the disciplines that required assimilation. Of course, the primary characteristic remains that of a professional armed with science and technology: there is a basic and unavoidable teaching, which already occupies a non-compressible part of the curriculum. However, the engineer cannot be a pure technician; he is called to become a manager, even to create a company: management courses, if not finance, are essential. Let us not forget CSR, corporate social responsibility. As such, but also because it is necessary to be attentive to society and more prosaically to perceive the expectations of the client, the emphasis is now placed on at least partial assimilation of the human and social sciences, the practical application of which to realize that they contain theories that must be assimilated, methodologies that must be applied and analyses of survey results that must be learned to interpret in order to overcome common sense, often misleading, etc. Special modules dedicated to innovation should be added. Should today's engineer become the 21st Century Pico della Mirandola?

Paradoxically, with regard to this broadly open range of disciplines, the question of the company's internal knowledge is not addressed. However, the engineer will not only work with their peers, but with all the social groups that make up the human fabric of the company's factory or offices. Observation and understanding of their modes of functioning will depend on the newcomer's ability to apprehend the collective work as a whole and the social relations that are established there, and thus to be able to integrate into this complex process, all the more so when, by virtue of their status, they are led to take on responsibilities. This was the problem developed by Bertrand Schwartz when he initiated the famous reform of the École des mines de

Nancy which, at least at the time, had pushed back the small world of the graduate engineering schools[10].

The relations with the future company, therefore the employer, are presented unequivocally as *a priori* harmonious, which makes sense from an educational perspective. The innovation processes studied in these texts require a climate of trust to develop and be experienced, and of benevolence in the face of risk taking, that is to say the possibility of failure. But firms have their own coherence that is not, in general, based on a generous altruism and a Fourierist conception of the social bond. Beginners as they may be, the engineer must not be naive or they shall face real disillusionment. It is one of the roles of the teaching teams to be able to draw attention to possible difficulties. Moreover, in real life, unlike their elders, sometimes paralyzed by their unwavering loyalty to their company, today's engineers have learned how to leave a company to better manage their own career.

One last point perhaps. The 1934 law confirmed that engineering degrees were not national degrees, such as the bachelor's or doctorate, but establishment degrees. So one is an engineer graduated from such a school before being a graduate engineer. It follows that school patriotism develops, particularly through alumni associations, which may tend to consider that the institution from which one comes as the best and that the others are less prestigious. This is the rationale that the 206 alumni associations are likely to develop for each of them. Such considerations undermine the notion of professional identity. Other elements also play against the perception of a professional body. The engineers mode of action, perfectly generalist at the beginning, did not stop, because of the development of the sciences, to be more specific. At the same time, we observe the indefinite extension of their field of intervention into areas where, in principle, we did not expect them (for example today we think of bioengineering in the field of nanotechnology). Given such an observation, the question of a common language for engineers can legitimately arise, which means that there is no longer an internal dialogue and it is no longer possible to determine collective issues. This is the myth of Babel revisited. Commonly, there remains only the title "engineer" and the title recognized by an administrative body for a collection of individuals. In these circumstances, would it still be lawful to speak of the engineering profession? If it is considered legitimate to resume the question again, it should be postulated

10 [BIR 98].

that there is potentially a collective memory of the engineers. In order to restore and revive it, the engineers would be well advised to take an interest in their own history and its characteristics, and to mobilize in order to cultivate it, develop it, diffuse it and then to find common values. Hence, we have an interest in knowing one's own history and characteristics.

References

[BAU 93] BAUDET J.C., "The training of engineers in Belgium, 1830-1940", in FOX R., GUAGNINI A. (eds), *Education, Technology and Industrial Performance in Europe, 1850-1939*, Cambridge University Press, Cambridge, pp. 93–114, 1993.

[BIR 98] BIRCK F., "L'École des mines de Nancy, de l'après-guerre à la réforme", interview with Bertrand Schwartz, in GRELON A., BIRCK F. (eds), *Des ingénieurs pour la Lorraine – XIXe-XXe siècle*, Éditions Serpenoise, Metz, pp. 217–235, 1998.

[GIL 64] GILLE B., *Les ingénieurs de la Renaissance*, Hermann, Paris, 1964.

[JAC 90] JACOMY B., *Une histoire des techniques*, Le Seuil, Paris, 1990.

[HIL 00] HILAIRE-PEREZ L., *L'invention technique au siècle des Lumières*, Albin Michel, Paris, 2000.

[LUS 04] LUSAMONFORTE G., "L'enseignement industriel pendant la première phase de l'industrialisation espagnole: l'École d'ingénieurs de Barcelone", in GOUZEVITCH I., GRELON A., KARVAR A. (eds), *La formation des ingénieurs en perspective. Modèles de référence et réseaux de médiation, XVIIIe-XXe siècle*, PUR, Rennes, pp. 35–52, 2004.

[MAR 86] MARSH J., "Du cercle privé à l'antichambre de l'Université: les associations d'ingénieurs et l'image de marque des ingénieurs britanniques du XVIIIe siècle à nos jours", in GRELON A. (ed.), *Les ingénieurs de la crise. Titre et profession entre les deux guerres*, Éditions de l'École des hautes études en sciences sociales, Paris, pp. 241–254, 1986.

[PAQ 04] PAQUIER S., "Les exemples contrastés de l'École d'ingénieurs de Lausanne et de l'École polytechnique fédérale de Zurich (1853–1914)", in GOUZEVITCH I., GRELON A., KARVAR A. (eds), *La formation des ingénieurs en perspective. Modèles de référence et réseaux de médiation, XVIIIe-XXe siècle*, PUR, Rennes, pp. 23–33, 2004.

[VIR 17] VIROL M., "La gloire d'un ingénieur. Intelligence technique et/ou stratégies d'écriture, Henri Gautier (Nîmes 1660-Paris 1737)", in CARVAIS R., GARÇON A.F., GRELON A. (eds), *Penser la technique autrement, XVIe—XXIe siècle. En hommage à l'œuvre d'Hélène Vérin*, Éditions Classiques Garnier, Paris, 2017.

List of Authors

Catherine ADAM
ENSTA Bretagne
Brest
France

Jane ANDREWS
Aston University
Birmingham
United Kingdom

Isabelle CAILLEAU
UTC
Sorbonne Universités
Compiègne, Paris
France

Emmanuel CARDONA GIL
ENSTA Bretagne
Brest
France

Hugues CHOPLIN
UTC
Sorbonne Universités
Compiègne, Paris
France

Robin CLARK
Aston University
Birmingham
United Kingdom

Serge COCO
CREAD
Rennes
France

Péter CSIZMADIA
Institute for Sociology of
Hungarian Academy
of Sciences
Budapest
Hungary

Linda GARDELLE
ENSTA Bretagne
Brest
France

Christiane GILLET
ENSTA Bretagne
Brest
France

André GRELON
EHESS
Paris
France

Tobias HAERTEL
Technical University of
Dortmund
Germany

Frédéric HUET
UTC
Sorbonne Universités
Compiègne, Paris
France

Anne-Marie JOLLY
Commission des Titres
d'Ingénieur
Paris
France

Klara KÖVESI
ENSTA Bretagne
Brest
France

Liudvika LEISYTE
Technical University of
Dortmund
Germany

Denis LEMAÎTRE
ENSTA Bretagne
Brest
France

Tiphaine LIU
ENS Paris-Saclay
Cachan
France

Dominik MAY
Technical University of
Dortmund
Germany

Christophe MORACE
ENSTA Bretagne
Brest
France

Julie NOLLAND
Commission des Titres
d'Ingénieur
Paris
France

Anna-Lena ROSE
Technical University of
Dortmund
Germany

Pierre STEINER
UTC
Sorbonne Universités
Compiègne, Paris
France

Brad TABAS
ENSTA Bretagne
Brest
France

Claudius TERKOWSKY
Technical University of Dortmund
Germany

Zsuzsanna VEROSZTA
Hungarian Demographic Research Institute
Budapest
Hungary

Index

A, B

C, D

E, G

F, H

I, L

M, O

P, R

S, U

Other titles from

iSTE

in

Innovation, Entrepreneurship and Management

2018

BURKHARDT Kirsten
Private Equity Firms: Their Role in the Formation of Strategic Alliances

CALLENS Stéphane
Creative Globalization
(Smart Innovation Set – Volume 16)

CASADELLA Vanessa
Innovation Systems in Emerging Economies: MINT – Mexico, Indonesia, Nigeria, Turkey
(Smart Innovation Set – Volume 18)

CHOUTEAU Marianne, FOREST Joëlle, NGUYEN Céline
Science, Technology and Innovation Culture
(Innovation in Engineering and Technology Set – Volume 3)

CORLOSQUET-HABART Marine, JANSSEN Jacques
Big Data for Insurance Companies
(Big Data, Artificial Intelligence and Data Analysis Set – Volume 1)

CROS Françoise
Innovation and Society
(Smart Innovation Set – Volume 15)

DEBREF Romain
Environmental Innovation and Ecodesign: Certainties and Controversies
(Smart Innovation Set – Volume 17)

DOMINGUEZ Noémie
SME Internationalization Strategies: Innovation to Conquer New Markets

ERMINE Jean-Louis
Knowledge Management: The Creative Loop
(Innovation and Technology Set – Volume 5)

GILBERT Patrick, BOBADILLA Natalia, GASTALDI Lise,
LE BOULAIRE Martine, LELEBINA Olga
Innovation, Research and Development Management

IBRAHIMI Mohammed
Mergers & Acquisitions : Theory, Strategy, Finance

LÉVY Aldo, BEN BOUHENI Faten, AMMI Chantal
Financial Management: USGAAP and IFRS Standards
(Innovation and Technology Set – Volume 6)

MILLOT Michel
Embarrassment of Product Choices 1: How to Consume Differently

SACHETTI Philippe, ZUPPINGER Thibaud
New Technologies and Branding
(Innovation and Technology Set – Volume 4)

TEMPLE Ludovic, COMPAORÉ SAWADOGO Eveline M.F.W.
Innovation Processes in Agro-Ecological Transitions in Developing Countries
(Innovation in Engineering and Technology Set – Volume 2)

UZUNIDIS Dimitri
Collective Innovation Processes: Principles and Practices
(Innovation in Engineering and Technology Set – Volume 4)

VAN HOOREBEKE Delphine
The Management of Living Beings or Emo-management

2017

AÏT-EL-HADJ Smaïl
The Ongoing Technological System
(Smart Innovation Set – Volume 11)

BAUDRY Marc, DUMONT Béatrice
Patents: Prompting or Restricting Innovation?
(Smart Innovation Set – Volume 12)

BÉRARD Céline, TEYSSIER Christine
Risk Management: Lever for SME Development and Stakeholder Value Creation

CHALENÇON Ludivine
Location Strategies and Value Creation of International Mergers and Acquisitions

CHAUVEL Danièle, BORZILLO Stefano
The Innovative Company: An Ill-defined Object
(Innovation Between Risk and Reward Set – Volume 1)

CORSI Patrick
Going Past Limits To Growth

D'ANDRIA Aude, GABARRET Inés
Building 21st Century Entrepreneurship
(Innovation and Technology Set – Volume 2)

DAIDJ Nabyla
Cooperation, Coopetition and Innovation
(Innovation and Technology Set – Volume 3)

FERNEZ-WALCH Sandrine
The Multiple Facets of Innovation Project Management
(Innovation between Risk and Reward Set – Volume 4)

FOREST Joëlle
Creative Rationality and Innovation
(Smart Innovation Set – Volume 14)

GUILHON Bernard
Innovation and Production Ecosystems
(Innovation between Risk and Reward Set – Volume 2)

HAMMOUDI Abdelhakim, DAIDJ Nabyla
Game Theory Approach to Managerial Strategies and Value Creation
(Diverse and Global Perspectives on Value Creation Set – Volume 3)

LALLEMENT Rémi
Intellectual Property and Innovation Protection: New Practices and New Policy Issues
(Innovation between Risk and Reward Set – Volume 3)

LAPERCHE Blandine
Enterprise Knowledge Capital
(Smart Innovation Set – Volume 13)

LEBERT Didier, EL YOUNSI Hafida
International Specialization Dynamics
(Smart Innovation Set – Volume 9)

MAESSCHALCK Marc
Reflexive Governance for Research and Innovative Knowledge
(Responsible Research and Innovation Set – Volume 6)

MASSOTTE Pierre
Ethics in Social Networking and Business 1: Theory, Practice and Current Recommendations
Ethics in Social Networking and Business 2: The Future and Changing Paradigms

MASSOTTE Pierre, CORSI Patrick
Smart Decisions in Complex Systems

MEDINA Mercedes, HERRERO Mónica, URGELLÉS Alicia
Current and Emerging Issues in the Audiovisual Industry
(Diverse and Global Perspectives on Value Creation Set – Volume 1)

MICHAUD Thomas
Innovation, Between Science and Science Fiction
(Smart Innovation Set – Volume 10)

PELLÉ Sophie
Business, Innovation and Responsibility
(Responsible Research and Innovation Set – Volume 7)

SAVIGNAC Emmanuelle
The Gamification of Work: The Use of Games in the Workplace

SUGAHARA Satoshi, DAIDJ Nabyla, USHIO Sumitaka
Value Creation in Management Accounting and Strategic Management: An Integrated Approach
(Diverse and Global Perspectives on Value Creation Set –Volume 2)

UZUNIDIS Dimitri, SAULAIS Pierre
Innovation Engines: Entrepreneurs and Enterprises in a Turbulent World
(Innovation in Engineering and Technology Set – Volume 1)

2016

BARBAROUX Pierre, ATTOUR Amel, SCHENK Eric
Knowledge Management and Innovation
(Smart Innovation Set – Volume 6)

BEN BOUHENI Faten, AMMI Chantal, LEVY Aldo
Banking Governance, Performance And Risk-Taking: Conventional Banks Vs Islamic Banks

BOUTILLIER Sophie, CARRÉ Denis, LEVRATTO Nadine
Entrepreneurial Ecosystems (Smart Innovation Set – Volume 2)

BOUTILLIER Sophie, UZUNIDIS Dimitri
The Entrepreneur (Smart Innovation Set – Volume 8)

BOUVARD Patricia, SUZANNE Hervé
Collective Intelligence Development in Business

GALLAUD Delphine, LAPERCHE Blandine
Circular Economy, Industrial Ecology and Short Supply Chains
(Smart Innovation Set – Volume 4)

GUERRIER Claudine
Security and Privacy in the Digital Era
(Innovation and Technology Set – Volume 1)

MEGHOUAR Hicham
Corporate Takeover Targets

MONINO Jean-Louis, SEDKAOUI Soraya
Big Data, Open Data and Data Development
(Smart Innovation Set – Volume 3)

MOREL Laure, LE ROUX Serge
Fab Labs: Innovative User
(Smart Innovation Set – Volume 5)

PICARD Fabienne, TANGUY Corinne
Innovations and Techno-ecological Transition
(Smart Innovation Set – Volume 7)

2015

CASADELLA Vanessa, LIU Zeting, DIMITRI Uzunidis
Innovation Capabilities and Economic Development in Open Economies
(Smart Innovation Set – Volume 1)

CORSI Patrick, MORIN Dominique
Sequencing Apple's DNA

CORSI Patrick, NEAU Erwan
Innovation Capability Maturity Model

FAIVRE-TAVIGNOT Bénédicte
Social Business and Base of the Pyramid

GODÉ Cécile
Team Coordination in Extreme Environments

MAILLARD Pierre
Competitive Quality and Innovation

MASSOTTE Pierre, CORSI Patrick
Operationalizing Sustainability

MASSOTTE Pierre, CORSI Patrick
Sustainability Calling

2014

DUBÉ Jean, LEGROS Diègo
Spatial Econometrics Using Microdata

LESCA Humbert, LESCA Nicolas
Strategic Decisions and Weak Signals

2013

HABART-CORLOSQUET Marine, JANSSEN Jacques, MANCA Raimondo
VaR Methodology for Non-Gaussian Finance

2012

DAL PONT Jean-Pierre
Process Engineering and Industrial Management

MAILLARD Pierre
Competitive Quality Strategies

POMEROL Jean-Charles
Decision-Making and Action

SZYLAR Christian
UCITS Handbook

2011

LESCA Nicolas
Environmental Scanning and Sustainable Development

LESCA Nicolas, LESCA Humbert
Weak Signals for Strategic Intelligence: Anticipation Tool for Managers

MERCIER-LAURENT Eunika
Innovation Ecosystems

2010

SZYLAR Christian
Risk Management under UCITS III/IV

2009

COHEN Corine
Business Intelligence

ZANINETTI Jean-Marc
Sustainable Development in the USA

2008

CORSI Patrick, DULIEU Mike
The Marketing of Technology Intensive Products and Services

DZEVER Sam, JAUSSAUD Jacques, ANDREOSSO Bernadette
Evolving Corporate Structures and Cultures in Asia: Impact of Globalization

2007

AMMI Chantal
Global Consumer Behavior

2006

BOUGHZALA Imed, ERMINE Jean-Louis
Trends in Enterprise Knowledge Management

CORSI Patrick *et al.*
Innovation Engineering: the Power of Intangible Networks

Printed and bound by CPI Group (UK) Ltd, Croydon, CR0 4YY

15/09/2026

14962351-0003